# SUDOKU

## 100 PUZZLES WITH SOLUTIONS

### MULTI LEVEL **BOOK 6**

ISBN: 9781694062949

A standard Sudoku puzzle consists of a grid of 9 blocks. Each block contains 9 boxes arranged in 3 rows and 3 columns.

| Column | | | Block | | | Box | | |
|---|---|---|---|---|---|---|---|---|
| 8 | 4 | 1 | 7 | 9 | 6 |   | 3 | 2 |
|   |   |   | 2 | 5 | 4 | 8 | 1 |   |
| 6 | 2 | 5 | 3 | 1 |   | 9 | 4 | 7 |
| 2 |   | 8 |   | 3 | 7 | 4 | 5 | 1 |
| 3 |   | 7 | 4 | 8 |   | 2 | 6 | 9 |
|   | 5 | 6 | 9 |   | 1 |   |   | 8 |
| 5 |   |   |   |   |   |   | 2 | 3 |
| 1 | 6 | 3 | 8 | 4 | 2 | 7 | 9 | 5 |
| 9 | 7 | 2 | 5 | 6 | 3 | 1 | 8 | 4 |

Row

**The Basic Rules of Sudoku:**

- There's only one solution to a Sudoku puzzle. A puzzle is considered solved when all 81 boxes contain numbers by following the Sudoku rules.
- When you start a game of Sudoku some blocks already have numbers (fewer number make a harder puzzle). These numbers cannot be changed.
- Each column must contain every number from 1 to 9 and no two numbers in the same column can be duplicated.
- Each row must contain every number from 1 to 9 and no two numbers in the same row can be duplicated.
- Each block must contain every number from 1 to 9 and no two numbers in the same block can be duplicated.

**Here's the solved puzzle:**

| 8 | 4 | 1 | 7 | 9 | 6 | 5 | 3 | 2 |
|---|---|---|---|---|---|---|---|---|
| 7 | 3 | 9 | 2 | 5 | 4 | 8 | 1 | 6 |
| 6 | 2 | 5 | 3 | 1 | 8 | 9 | 4 | 7 |
| 2 | 9 | 8 | 6 | 3 | 7 | 4 | 5 | 1 |
| 3 | 1 | 7 | 4 | 8 | 5 | 2 | 6 | 9 |
| 4 | 5 | 6 | 9 | 2 | 1 | 3 | 7 | 8 |
| 5 | 8 | 4 | 1 | 7 | 9 | 6 | 2 | 3 |
| 1 | 6 | 3 | 8 | 4 | 2 | 7 | 9 | 5 |
| 9 | 7 | 2 | 5 | 6 | 3 | 1 | 8 | 4 |

## Puzzle 1
### Easy

| 7 | 4 | 2 |   | 3 | 9 | 6 |   |   |
|---|---|---|---|---|---|---|---|---|
|   | 9 |   | 4 | 6 | 2 | 8 | 7 | 3 |
| 8 | 6 | 3 |   |   | 7 | 2 |   | 9 |
| 5 | 2 |   | 3 | 7 | 8 | 9 | 1 | 6 |
| 3 | 8 | 1 | 5 | 9 | 6 | 7 | 2 | 4 |
| 6 | 7 | 9 | 2 | 1 | 4 |   | 3 | 8 |
| 2 | 5 |   | 9 | 4 | 3 | 1 | 6 |   |
| 4 | 1 |   |   | 8 |   |   | 9 | 2 |
|   | 3 |   |   | 2 | 1 | 4 | 8 | 5 |

## Puzzle 2
### Easy

| 8 | 9 | 5 |   | 7 |   |   | 4 | 2 |
|---|---|---|---|---|---|---|---|---|
| 7 | 4 | 3 |   | 8 | 2 |   | 9 |   |
| 6 | 2 | 1 | 4 | 5 | 9 | 3 |   | 7 |
| 4 | 6 | 2 | 9 | 3 | 5 | 7 | 1 | 8 |
| 5 |   |   | 8 |   | 6 | 9 | 2 |   |
|   | 8 |   | 2 |   | 7 | 5 |   |   |
|   |   |   | 5 |   | 8 | 2 | 7 | 9 |
| 2 | 7 | 6 |   | 9 | 4 | 8 | 5 | 1 |
| 9 | 5 | 8 | 7 | 2 | 1 | 4 | 3 | 6 |

## Puzzle 3
### Easy

| 4 | 5 |   |   | 3 | 1 | 8 | 2 | 7 |
|---|---|---|---|---|---|---|---|---|
| 7 |   |   | 9 | 5 | 2 | 4 | 1 | 3 |
| 3 |   | 2 |   | 7 | 4 | 5 |   |   |
| 1 | 3 | 4 |   | 2 | 5 |   |   | 8 |
| 2 |   | 7 |   | 6 | 8 | 3 | 5 |   |
| 5 | 8 |   | 3 | 4 | 9 | 2 | 7 | 1 |
| 8 | 7 | 5 | 2 | 1 | 3 |   | 4 |   |
| 9 | 4 | 1 | 5 | 8 | 6 | 7 | 3 | 2 |
| 6 | 2 | 3 | 4 | 9 | 7 | 1 | 8 | 5 |

## Puzzle 4
### Easy

|   | 3 | 5 | 1 | 8 |   | 4 | 9 | 7 |
|---|---|---|---|---|---|---|---|---|
| 4 | 8 |   |   |   |   | 6 | 1 | 3 |
| 1 | 6 | 9 | 4 | 3 | 7 | 8 | 5 | 2 |
| 8 | 7 | 6 |   |   | 1 | 3 | 2 | 5 |
| 5 |   | 4 | 3 | 7 | 2 |   | 6 | 8 |
| 3 | 2 | 1 | 6 | 5 | 8 | 9 | 7 | 4 |
| 7 |   | 3 |   |   |   |   | 8 | 6 |
| 6 |   | 8 | 7 |   |   |   | 3 | 9 |
| 9 | 5 | 2 | 8 | 6 | 3 | 7 | 4 | 1 |

## Puzzle 5
### Easy

| 9 | 8 |   |   | 4 |   | 2 | 3 | 6 |
|---|---|---|---|---|---|---|---|---|
| 2 | 5 | 3 | 6 | 8 | 9 | 4 | 7 | 1 |
| 4 |   |   | 3 | 2 |   | 8 | 5 | 9 |
|   | 4 | 5 |   | 9 | 2 | 6 |   |   |
| 6 | 1 | 2 |   | 3 |   | 9 | 8 |   |
|   | 3 | 9 | 4 | 1 | 6 | 5 | 2 | 7 |
| 1 |   | 8 | 9 | 6 | 3 | 7 | 4 | 5 |
| 3 |   |   | 2 | 5 |   | 1 | 9 | 8 |
|   | 9 |   | 1 | 7 | 8 | 3 | 6 | 2 |

## Puzzle 6
### Easy

| 3 | 8 | 9 | 2 | 7 | 6 | 1 | 4 | 5 |
|---|---|---|---|---|---|---|---|---|
|   | 2 | 6 | 3 | 5 | 1 | 9 | 8 | 7 |
| 5 |   | 7 |   | 8 | 9 | 2 | 6 | 3 |
|   | 4 | 1 | 8 | 3 |   | 5 | 2 | 9 |
| 2 | 5 | 3 |   | 1 | 4 | 8 | 7 | 6 |
| 7 |   | 8 | 5 | 6 | 2 |   |   | 1 |
|   |   | 4 | 7 | 2 | 5 | 6 |   | 8 |
|   | 7 | 5 |   | 4 | 8 | 3 | 1 | 2 |
| 8 | 6 |   |   |   |   | 7 | 5 | 4 |

## Puzzle 7
### Easy

| 7 |   |   | 5 | 2 | 4 |   | 3 |   |
|---|---|---|---|---|---|---|---|---|
| 3 | 1 | 2 | 9 | 6 | 8 | 4 | 7 |   |
| 6 | 4 | 5 | 7 | 3 | 1 | 8 | 2 | 9 |
| 5 | 7 |   | 6 | 8 | 9 | 2 | 1 |   |
| 1 | 2 | 6 | 3 | 4 | 7 | 9 | 5 | 8 |
|   |   |   |   | 1 | 5 |   | 6 |   |
| 2 | 5 | 1 | 4 | 9 | 6 |   | 8 |   |
| 4 | 3 | 7 | 8 | 5 | 2 |   | 9 |   |
|   | 6 |   | 1 | 7 | 3 | 5 | 4 | 2 |

## Puzzle 8
### Easy

| 3 |   | 5 | 9 | 7 |   | 8 | 1 | 4 |
|---|---|---|---|---|---|---|---|---|
| 1 | 8 | 9 | 5 | 4 |   |   | 3 | 6 |
| 7 | 6 | 4 |   | 3 | 8 | 9 | 2 | 5 |
| 6 |   | 7 | 2 |   | 3 | 5 | 8 | 1 |
|   | 5 |   | 7 | 6 |   | 3 | 4 |   |
| 9 | 3 |   | 4 | 8 | 5 |   | 7 | 2 |
| 4 | 1 |   | 3 | 5 | 9 | 2 |   | 7 |
| 5 | 7 |   |   | 2 | 4 | 1 | 9 | 3 |
| 2 | 9 | 3 | 6 | 1 | 7 | 4 | 5 | 8 |

## Puzzle 9
### Easy

| 8 | 4 | 6 |   | 7 | 3 | 2 |   | 5 |
|---|---|---|---|---|---|---|---|---|
|   | 1 |   | 6 | 2 | 5 | 7 |   | 8 |
| 5 | 2 | 7 |   | 4 | 8 |   |   | 3 |
| 4 |   |   | 2 | 6 | 1 |   | 5 | 7 |
| 1 | 5 | 2 | 4 | 8 |   | 3 | 6 | 9 |
| 7 | 6 | 8 | 5 | 3 | 9 | 4 | 2 | 1 |
|   |   | 4 |   | 1 | 6 | 5 |   | 2 |
| 6 |   | 5 |   | 9 | 2 | 1 |   | 4 |
| 2 | 8 | 1 | 7 | 5 | 4 | 9 | 3 | 6 |

## Puzzle 10
### Easy

| 2 | 3 |   |   | 6 | 1 | 7 | 8 | 4 |
|---|---|---|---|---|---|---|---|---|
|   | 7 |   |   | 4 |   | 6 |   | 1 |
| 1 | 4 | 6 |   |   | 8 | 9 |   | 2 |
|   | 2 |   | 8 |   | 5 | 1 | 7 | 3 |
| 3 | 5 | 1 | 7 | 2 | 4 | 8 |   | 9 |
| 9 | 8 | 7 | 1 | 3 |   | 4 | 2 | 5 |
| 7 | 1 | 2 | 4 | 8 | 3 | 5 | 9 | 6 |
| 4 |   | 3 |   | 5 | 7 | 2 | 1 | 8 |
| 5 |   | 8 |   | 1 |   | 3 | 4 | 7 |

## Puzzle 11
### Easy

| 6 |   | 7 |   | 5 |   | 2 | 3 | 1 |
|---|---|---|---|---|---|---|---|---|
| 3 | 9 | 4 |   | 1 |   | 8 | 5 | 7 |
| 2 | 1 |   | 7 | 8 | 3 | 9 | 4 |   |
| 8 |   | 3 |   | 2 | 5 |   | 9 | 4 |
| 9 |   | 2 |   | 3 | 8 |   | 7 | 5 |
| 5 | 6 | 1 | 9 | 4 | 7 |   | 2 |   |
| 1 | 2 | 9 | 8 |   | 4 | 5 | 6 |   |
| 4 | 3 | 8 | 5 | 6 | 2 | 7 | 1 | 9 |
| 7 | 5 | 6 | 3 | 9 | 1 | 4 | 8 | 2 |

## Puzzle 12
### Easy

|   |   |   | 9 | 3 | 2 | 4 | 7 | 1 |
|---|---|---|---|---|---|---|---|---|
| 4 | 3 | 9 | 7 | 1 | 5 | 6 | 2 | 8 |
| 7 | 1 | 2 | 8 | 6 | 4 | 5 | 3 | 9 |
|   |   |   | 5 | 7 | 3 | 1 |   | 4 |
|   |   |   | 2 | 4 | 8 | 3 |   | 7 |
| 3 | 7 | 4 | 1 | 9 | 6 | 8 | 5 | 2 |
|   |   |   | 4 | 8 | 9 | 7 | 1 |   |
|   |   | 7 | 3 | 5 | 1 | 2 | 4 | 6 |
|   | 4 |   | 6 | 2 | 7 | 9 | 8 |   |

## Puzzle 13
### Easy

| | 4 | 7 | 2 | 1 | | | | 5 |
|---|---|---|---|---|---|---|---|---|
| 3 | 1 | 5 | 7 | 8 | 6 | 2 | 9 | 4 |
| | 6 | 2 | 4 | 5 | | 1 | 7 | |
| 5 | 2 | 1 | 9 | 7 | | | | |
| 6 | 8 | 4 | 3 | 2 | 5 | 7 | 1 | 9 |
| 7 | 3 | 9 | 1 | 6 | | | | 2 |
| 2 | 9 | 3 | 6 | 4 | 7 | | | 1 |
| 1 | 5 | 6 | 8 | 9 | 2 | | | 7 |
| 4 | 7 | 8 | 5 | 3 | 1 | 9 | 2 | 6 |

## Puzzle 14
### Easy

| | 1 | 6 | 4 | | 5 | | 7 | 8 |
|---|---|---|---|---|---|---|---|---|
| 8 | 5 | 9 | 6 | | | 4 | | 1 |
| 4 | 7 | | 9 | 1 | 8 | 6 | | |
| 1 | 8 | 5 | 7 | 4 | 6 | 3 | 9 | 2 |
| 9 | | 7 | 2 | 5 | 3 | 1 | | 4 |
| | 4 | | 8 | 9 | 1 | | 6 | 7 |
| 7 | | | 1 | 6 | 9 | 8 | | |
| 5 | | 8 | 3 | 7 | 4 | | 1 | 6 |
| 6 | 3 | 1 | 5 | 8 | 2 | 7 | 4 | 9 |

## Puzzle 15
### Easy

| 4 | 2 | 6 | 3 | 1 |   |   | 7 |   |
|---|---|---|---|---|---|---|---|---|
| 8 |   | 1 | 4 | 7 |   | 2 | 6 | 3 |
| 7 |   | 3 | 2 | 8 |   |   |   |   |
| 3 | 4 | 9 | 7 | 2 | 8 | 6 | 5 | 1 |
| 2 | 7 | 5 | 1 | 6 | 3 | 9 | 8 | 4 |
| 6 | 1 | 8 |   |   | 4 | 3 | 2 | 7 |
| 9 | 3 | 7 | 6 |   | 2 |   |   | 8 |
| 5 | 6 | 4 | 8 | 3 | 1 | 7 | 9 | 2 |
| 1 | 8 | 2 |   |   | 7 |   | 3 | 6 |

## Puzzle 16
### Easy

| 4 | 5 | 2 |   | 6 | 9 | 1 | 8 | 3 |
|---|---|---|---|---|---|---|---|---|
| 3 |   | 9 | 8 | 1 | 4 |   |   | 6 |
| 1 | 6 | 8 | 2 | 3 | 5 | 4 | 9 | 7 |
| 5 | 9 | 1 | 3 | 4 | 2 | 7 | 6 | 8 |
| 6 | 8 |   | 9 | 5 | 1 |   |   | 4 |
| 2 |   | 4 | 6 | 8 | 7 |   | 1 |   |
| 8 | 1 | 5 | 4 |   |   | 6 | 7 | 2 |
| 9 | 2 | 3 | 5 |   | 6 | 8 | 4 | 1 |
| 7 | 4 |   | 1 | 2 | 8 |   |   |   |

## Puzzle 17
### Easy

| 4 | 5 | 1 | 2 |   | 7 | 6 | 8 |   |
|---|---|---|---|---|---|---|---|---|
| 2 | 8 | 9 | 6 | 3 | 1 | 5 | 4 | 7 |
| 6 | 7 | 3 | 5 |   |   |   | 1 | 2 |
| 5 |   | 4 | 7 | 1 | 3 | 8 |   | 6 |
| 1 |   | 2 | 4 |   |   | 7 | 3 | 9 |
| 8 | 3 | 7 | 9 | 6 | 2 | 1 |   | 4 |
| 3 | 2 |   |   | 7 |   | 4 | 6 | 5 |
|   | 1 | 5 | 3 | 4 | 6 |   | 9 | 8 |
| 9 |   |   |   | 2 |   | 3 | 7 | 1 |

## Puzzle 18
### Easy

| 4 | 6 | 9 | 2 | 1 | 8 | 3 | 7 | 5 |
|---|---|---|---|---|---|---|---|---|
| 3 | 7 |   | 6 |   | 9 | 2 |   | 8 |
| 2 | 8 |   | 3 | 7 |   | 6 |   |   |
| 8 | 1 | 6 | 4 | 9 | 7 | 5 | 3 | 2 |
| 9 | 5 | 4 | 1 |   | 3 | 7 | 8 | 6 |
| 7 | 3 |   |   |   | 6 |   |   | 1 |
| 5 |   | 8 | 7 | 3 |   |   | 6 |   |
| 1 | 4 | 7 | 9 | 6 | 2 | 8 | 5 | 3 |
| 6 | 9 | 3 |   |   |   |   | 2 | 7 |

## Puzzle 19
### Easy

| 6 | 8 | 9 | 1 | 3 | 4 | 7 | 5 | 2 |
|---|---|---|---|---|---|---|---|---|
|   |   |   | 8 | 6 | 5 | 3 | 1 | 9 |
| 5 | 1 | 3 | 9 | 7 | 2 |   |   | 8 |
| 1 | 9 |   |   | 2 | 7 |   | 8 |   |
|   |   |   |   | 9 | 8 |   |   | 1 |
| 3 | 5 | 8 | 4 | 1 | 6 | 2 | 9 | 7 |
| 7 | 6 | 1 | 2 | 5 | 9 | 8 |   |   |
| 9 | 4 | 5 | 7 | 8 | 3 | 1 | 2 | 6 |
| 8 | 3 | 2 | 6 | 4 | 1 | 9 | 7 | 5 |

## Puzzle 20
### Easy

| 3 |   | 5 | 9 | 6 | 8 | 4 | 1 | 7 |
|---|---|---|---|---|---|---|---|---|
|   | 6 | 1 |   | 7 |   | 2 | 5 | 9 |
| 9 | 7 | 4 | 5 | 2 | 1 | 3 | 6 | 8 |
| 1 | 9 | 2 |   |   |   | 7 | 8 | 5 |
| 6 | 4 |   | 2 | 5 | 7 | 1 | 9 | 3 |
| 7 | 5 | 3 | 1 |   | 9 | 6 |   | 4 |
|   | 1 | 9 |   |   | 5 | 8 | 7 |   |
|   | 8 | 6 | 7 | 9 |   | 5 | 3 | 1 |
| 5 | 3 | 7 | 8 | 1 |   | 9 | 4 |   |

## Puzzle 21
### Easy

| 7 | 1 | 9 | 2 | 4 | 8 | 3 | 5 | 6 |
|---|---|---|---|---|---|---|---|---|
| 8 | 5 | 3 | 1 | 6 | 7 | 2 |   | 9 |
| 6 | 4 |   | 9 | 3 | 5 | 7 |   |   |
| 5 |   |   | 8 | 2 | 4 | 1 | 9 | 3 |
| 9 | 2 |   | 5 | 7 | 3 | 8 | 6 | 4 |
| 4 | 3 | 8 | 6 | 1 | 9 | 5 | 7 | 2 |
| 1 | 9 | 5 | 4 | 8 | 2 | 6 | 3 | 7 |
| 2 |   |   | 3 |   |   |   |   |   |
| 3 |   |   | 7 |   |   |   | 2 |   |

## Puzzle 22
### Easy

| 6 |   | 4 | 7 | 5 | 9 | 3 | 1 | 8 |
|---|---|---|---|---|---|---|---|---|
| 7 | 1 | 8 |   |   |   | 9 | 5 | 2 |
| 3 |   | 9 | 8 |   | 1 | 4 | 6 | 7 |
| 5 | 7 |   |   |   |   | 8 | 9 |   |
| 4 |   |   |   | 9 | 5 | 7 | 3 |   |
|   | 6 | 3 | 1 | 8 | 7 | 2 | 4 | 5 |
| 2 | 3 | 5 | 9 | 7 | 6 | 1 | 8 | 4 |
| 1 | 9 | 7 | 5 | 4 |   |   | 2 | 3 |
| 8 | 4 | 6 | 3 | 1 | 2 | 5 | 7 | 9 |

## Puzzle 23
### Easy

| 7 | 4 |   | 5 | 6 | 9 | 1 | 3 | 2 |
|---|---|---|---|---|---|---|---|---|
| 3 | 1 | 9 | 2 | 4 | 7 |   | 8 |   |
| 5 |   | 6 | 1 | 8 | 3 | 9 | 4 | 7 |
|   |   | 1 | 7 | 9 | 4 | 2 | 5 | 3 |
| 9 | 3 | 4 | 8 | 2 | 5 | 7 | 6 | 1 |
| 2 | 7 | 5 | 3 | 1 | 6 | 8 |   | 4 |
| 4 |   | 3 |   | 7 | 1 |   | 2 |   |
|   |   | 7 | 4 | 5 | 2 | 3 | 1 |   |
| 1 |   | 2 |   | 3 | 8 | 4 | 7 |   |

## Puzzle 24
### Easy

|   | 4 | 9 | 5 | 7 | 6 | 1 |   | 8 |
|---|---|---|---|---|---|---|---|---|
| 7 |   | 1 | 4 | 2 | 8 | 9 | 5 |   |
|   |   |   | 1 |   | 9 | 4 | 7 | 6 |
|   | 3 |   | 6 | 4 | 5 | 2 | 8 | 9 |
|   |   | 6 | 3 |   | 1 | 5 | 4 | 7 |
| 9 | 5 | 4 | 2 | 8 | 7 | 6 | 3 | 1 |
| 6 | 7 | 8 | 9 |   | 4 | 3 | 1 |   |
| 5 | 9 | 2 | 7 | 1 | 3 | 8 | 6 | 4 |
|   |   | 3 |   |   |   |   | 9 | 5 |

## Puzzle 25
### Easy

| 8 | 5 | 1 | 4 | 9 | 3 | 6 | 7 | 2 |
|---|---|---|---|---|---|---|---|---|
|   |   | 4 | 2 | 1 | 6 | 8 | 5 | 9 |
| 6 | 9 | 2 | 8 | 5 | 7 | 3 | 4 | 1 |
| 5 | 1 | 9 | 3 | 8 | 2 | 7 | 6 | 4 |
| 4 | 6 | 8 | 1 | 7 | 9 | 5 | 2 | 3 |
|   |   | 7 | 6 | 4 | 5 | 9 | 1 | 8 |
|   |   | 3 |   |   | 4 | 1 | 8 |   |
| 1 | 4 | 5 |   |   | 8 | 2 |   |   |
|   | 8 | 6 |   |   | 1 | 4 |   |   |

## Puzzle 26
### Medium

|   |   | 1 | 3 | 2 | 4 | 7 |   | 5 |
|---|---|---|---|---|---|---|---|---|
| 7 | 5 |   |   |   | 8 | 1 | 2 |   |
| 3 | 2 | 6 | 7 | 1 | 5 | 8 | 4 | 9 |
| 1 |   |   |   | 5 | 3 |   |   |   |
| 6 | 8 | 9 | 2 |   | 7 |   |   | 1 |
|   |   | 5 |   | 8 | 6 |   | 9 |   |
| 4 |   | 8 | 5 |   |   |   | 1 |   |
| 9 | 1 | 3 | 8 |   | 2 |   | 7 | 4 |
| 5 | 6 | 2 | 4 | 7 | 1 | 9 | 3 | 8 |

## Puzzle 27
## Medium

| | | | | | | | 3 | 6 |
|---|---|---|---|---|---|---|---|---|
| 3 | | | 6 | | | 7 | 1 | |
| | 9 | 6 | | | 3 | | 8 | |
| | 3 | 7 | 1 | 4 | 6 | | 2 | |
| 9 | | 5 | 2 | 3 | 7 | | | 8 |
| 6 | | 2 | 8 | 5 | 9 | 3 | 7 | 1 |
| 4 | 6 | 9 | 3 | 7 | | 8 | 5 | 2 |
| 5 | 7 | 1 | 9 | 8 | 2 | | | 3 |
| 2 | 8 | 3 | 5 | 6 | 4 | 1 | 9 | 7 |

## Puzzle 28
## Medium

| 6 | | 5 | 3 | 8 | 7 | 1 | 4 | 2 |
|---|---|---|---|---|---|---|---|---|
| 3 | 2 | | 5 | 6 | 4 | 9 | 8 | 7 |
| | | | 9 | 1 | 2 | 6 | 5 | 3 |
| 8 | 1 | 3 | 2 | 7 | 9 | 5 | 6 | 4 |
| 9 | 6 | 2 | | | | 7 | 3 | 1 |
| | | 7 | 6 | 3 | 1 | 8 | 2 | 9 |
| | | 9 | 7 | | | | 1 | |
| | | | 1 | | | | 9 | |
| 1 | | | | 9 | | 2 | | |

## Puzzle 29
## Medium

| 2 |   | 4 |   | 6 | 8 | 1 | 9 | 5 |
|---|---|---|---|---|---|---|---|---|
| 5 | 6 | 1 | 9 | 4 | 2 | 3 | 8 | 7 |
|   |   |   |   |   |   | 4 | 2 | 6 |
| 7 | 8 | 6 | 2 | 9 | 3 | 5 | 1 | 4 |
| 9 |   | 3 | 6 |   |   | 8 | 7 | 2 |
|   |   | 2 |   | 8 |   | 6 | 3 | 9 |
|   |   | 8 |   |   |   | 2 | 5 | 3 |
| 3 |   | 5 | 8 | 2 |   | 7 |   | 1 |
|   | 2 | 7 |   |   |   | 9 | 4 | 8 |

## Puzzle 30
## Medium

| 1 | 4 | 8 |   |   |   |   | 5 |   |
|---|---|---|---|---|---|---|---|---|
| 9 |   | 5 | 4 | 8 |   | 2 | 1 | 7 |
| 7 | 6 | 2 | 1 | 9 | 5 | 3 | 4 | 8 |
| 6 | 1 | 9 | 8 | 3 | 2 |   | 7 | 5 |
|   |   |   | 6 | 4 | 9 | 1 | 8 | 2 |
| 8 | 2 | 4 | 5 | 1 | 7 |   |   | 3 |
|   |   |   |   |   | 4 | 8 |   |   |
| 4 |   |   |   |   | 8 | 7 |   |   |
| 2 | 8 |   | 9 |   | 1 | 5 | 3 | 4 |

## Puzzle 31
## Medium

| 7 | 6 |   | 4 | 2 |   |   |   |   |
|---|---|---|---|---|---|---|---|---|
| 9 | 8 |   | 7 | 6 |   |   | 4 |   |
| 4 | 3 |   | 8 |   |   |   |   |   |
| 1 |   | 7 | 3 | 8 | 2 |   |   |   |
| 8 |   | 9 | 6 | 5 | 1 |   |   |   |
| 3 |   | 6 | 9 | 4 | 7 |   |   | 1 |
| 5 | 7 | 4 | 2 | 3 | 8 | 6 | 1 | 9 |
| 2 | 9 | 8 | 1 | 7 |   |   | 5 |   |
| 6 | 1 | 3 | 5 | 9 | 4 | 8 | 7 | 2 |

## Puzzle 32
## Medium

|   | 4 | 1 |   |   | 6 |   | 2 | 7 |
|---|---|---|---|---|---|---|---|---|
| 7 | 6 |   | 8 | 3 | 1 | 4 | 5 | 9 |
| 8 |   | 9 | 4 |   | 2 | 6 | 1 |   |
|   | 8 |   | 7 | 1 | 9 | 3 | 4 | 2 |
|   |   |   |   |   |   |   |   | 8 |
| 9 | 3 | 7 | 2 | 4 |   | 5 | 6 |   |
| 2 | 7 | 3 |   |   | 4 | 1 |   | 5 |
| 4 | 9 | 8 | 1 | 2 | 5 | 7 | 3 | 6 |
|   | 1 | 5 | 3 | 8 | 7 |   |   | 4 |

## Puzzle 33
### Medium

| 3 | 5 | 7 | 2 | 1 | 4 | 8 |   | 9 |
|---|---|---|---|---|---|---|---|---|
| 1 | 6 | 4 | 5 | 8 | 9 | 2 | 7 | 3 |
|   | 8 |   | 6 |   | 3 | 4 | 5 | 1 |
|   | 4 | 8 | 7 | 9 | 2 |   |   | 5 |
|   |   |   | 3 |   |   | 7 | 8 | 4 |
| 5 | 7 | 3 | 8 |   | 1 | 6 | 9 | 2 |
| 4 |   | 6 |   | 5 | 7 |   |   | 8 |
| 8 |   |   |   |   |   | 5 |   | 7 |
| 7 |   | 5 |   |   | 8 |   |   | 6 |

## Puzzle 34
### Medium

|   | 5 | 3 | 1 | 9 |   | 2 | 4 | 7 |
|---|---|---|---|---|---|---|---|---|
| 8 |   |   |   | 7 | 4 | 6 |   | 5 |
| 4 | 2 | 7 | 5 | 3 | 6 |   | 8 |   |
| 7 | 3 | 1 | 9 | 8 | 5 | 4 | 6 | 2 |
| 5 |   | 6 | 4 | 1 | 2 | 3 | 7 |   |
| 2 | 4 | 8 | 3 |   | 7 | 5 |   | 9 |
|   |   | 2 | 8 | 5 | 1 | 7 | 9 | 4 |
|   | 7 |   | 6 | 4 | 9 |   |   | 3 |
|   |   |   |   |   |   |   |   |   |

## Puzzle 35
## Medium

| 5 |   | 9 | 4 |   | 8 | 2 |   | 3 |
|---|---|---|---|---|---|---|---|---|
| 6 | 2 | 4 |   |   | 3 |   | 8 |   |
|   | 7 | 3 |   |   | 2 |   | 1 | 4 |
| 3 | 6 |   |   |   | 9 | 4 |   | 7 |
| 9 | 8 | 2 | 7 | 4 | 6 | 5 | 3 | 1 |
| 1 | 4 | 7 |   |   | 5 | 8 | 9 | 6 |
| 4 | 3 |   |   |   | 1 |   |   |   |
| 7 | 9 | 1 |   |   | 4 | 3 | 5 |   |
| 2 | 5 |   |   |   | 7 | 1 | 4 | 9 |

## Puzzle 36
## Medium

| 1 | 2 |   | 3 | 8 | 4 | 6 |   | 5 |
|---|---|---|---|---|---|---|---|---|
| 5 | 4 | 8 | 9 | 7 | 6 | 2 | 3 | 1 |
| 6 | 3 |   | 5 | 1 | 2 | 8 |   | 4 |
|   |   |   | 2 | 9 | 1 | 3 |   | 6 |
| 9 | 6 | 2 |   | 5 | 3 |   | 1 | 8 |
| 3 |   |   |   | 6 |   |   | 2 | 9 |
|   |   | 6 | 1 | 3 | 5 |   | 8 | 2 |
| 2 |   |   | 8 | 4 | 7 |   |   | 3 |
|   |   | 3 | 6 | 2 | 9 |   |   | 7 |

## Puzzle 37
### Medium

| 3 | 4 | 6 | 7 | 5 |   | 1 | 9 |   |
|---|---|---|---|---|---|---|---|---|
| 8 |   | 1 | 9 | 3 |   |   | 5 |   |
| 5 | 7 |   | 4 |   | 2 | 3 |   | 6 |
|   |   |   | 1 |   |   |   | 4 | 3 |
|   |   | 3 | 6 | 7 | 9 | 2 | 1 |   |
|   |   |   | 8 | 4 | 3 |   | 7 | 9 |
|   | 9 | 5 | 3 | 6 | 4 |   | 2 |   |
| 1 |   | 4 | 2 | 8 | 7 | 9 |   | 5 |
| 2 |   | 8 | 5 | 9 | 1 | 4 |   |   |

## Puzzle 38
### Medium

|   | 3 |   |   |   | 2 | 1 | 5 | 9 |
|---|---|---|---|---|---|---|---|---|
| 1 | 8 | 5 | 9 | 6 | 3 | 7 | 2 | 4 |
| 9 | 2 | 4 |   |   | 7 | 3 | 6 | 8 |
| 8 |   | 1 | 7 |   | 6 | 4 |   | 2 |
| 2 |   | 3 |   |   |   | 6 |   | 7 |
|   |   |   | 2 |   |   | 5 |   | 1 |
| 5 | 7 | 8 | 3 |   |   | 2 | 4 | 6 |
| 4 | 6 | 2 |   | 7 |   | 9 | 1 | 3 |
| 3 |   |   | 6 | 2 | 4 | 8 | 7 | 5 |

## Puzzle 39
### Medium

| 4 | 9 |   | 3 |   | 8 | 2 | 5 |   |
|---|---|---|---|---|---|---|---|---|
| 5 |   |   |   | 6 | 2 |   | 8 | 4 |
| 6 |   |   | 9 | 4 |   |   | 7 |   |
| 1 | 6 | 2 |   | 7 |   |   |   | 5 |
| 8 |   |   | 1 | 9 |   |   | 2 | 7 |
| 7 | 4 | 9 | 5 | 2 | 6 |   | 1 |   |
|   | 7 | 5 |   |   | 9 | 8 |   | 1 |
| 3 | 8 | 4 | 2 | 5 | 1 | 7 |   | 9 |
| 9 | 1 | 6 | 4 | 8 | 7 | 5 | 3 | 2 |

## Puzzle 40
### Medium

| 5 |   | 6 |   | 3 | 1 | 8 | 2 | 9 |
|---|---|---|---|---|---|---|---|---|
| 2 | 1 | 3 | 9 | 8 | 6 | 4 | 7 | 5 |
| 9 | 8 |   |   |   |   | 6 | 1 | 3 |
| 6 |   | 1 |   | 2 | 3 | 5 |   |   |
|   |   | 8 |   |   |   | 2 |   |   |
| 4 |   | 2 |   |   | 7 |   |   |   |
| 8 | 2 | 4 |   | 9 |   | 7 | 3 | 6 |
| 1 | 6 | 5 | 3 | 7 | 4 | 9 | 8 | 2 |
| 7 | 3 | 9 | 2 | 6 | 8 | 1 |   |   |

## Puzzle 41
## Medium

| 9 | 2 |   |   |   | 5 | 3 | 8 | 7 |
|---|---|---|---|---|---|---|---|---|
| 8 | 4 | 7 | 3 | 9 | 2 |   |   | 6 |
|   |   |   | 7 |   |   | 9 | 2 | 4 |
|   |   |   | 9 | 2 | 3 |   | 6 | 8 |
|   |   |   | 8 |   | 6 |   | 3 |   |
| 3 | 6 | 8 | 5 | 7 | 1 | 2 | 4 |   |
| 4 |   |   | 2 | 8 | 7 | 6 |   |   |
|   | 8 |   | 1 | 5 | 9 | 4 | 7 | 3 |
| 1 | 7 | 5 | 6 | 3 | 4 | 8 | 9 | 2 |

## Puzzle 42
## Medium

| 3 |   |   |   | 6 |   | 1 |   | 7 |
|---|---|---|---|---|---|---|---|---|
| 5 |   |   |   | 2 |   | 3 |   | 4 |
| 8 |   |   | 3 |   |   | 6 |   | 5 |
| 6 |   |   |   | 4 |   | 8 | 3 | 9 |
|   |   | 5 |   | 3 | 8 |   | 7 | 1 |
| 7 | 8 | 3 | 2 | 1 | 9 |   | 5 | 6 |
| 2 | 5 | 8 | 4 | 7 | 6 | 9 | 1 | 3 |
| 9 | 3 | 6 |   | 5 | 2 | 7 | 4 |   |
| 1 |   |   | 9 | 8 | 3 | 5 | 6 | 2 |

## Puzzle 43
## Medium

| 3 |   |   |   |   |   |   |   | 6 |
|---|---|---|---|---|---|---|---|---|
| 1 | 5 | 2 | 6 | 3 | 4 |   | 8 | 7 |
| 6 |   |   |   |   |   |   |   |   |
| 7 | 6 | 4 | 5 | 1 | 3 | 2 | 9 | 8 |
| 9 | 2 | 5 | 7 | 4 | 8 | 6 | 1 | 3 |
| 8 | 3 | 1 | 2 | 6 |   |   |   |   |
| 2 | 7 | 8 |   | 9 | 6 |   | 5 |   |
|   |   | 6 |   | 8 | 2 |   |   | 9 |
| 4 | 9 | 3 | 1 | 5 | 7 | 8 | 6 | 2 |

## Puzzle 44
## Medium

| 1 |   | 6 |   | 2 |   | 5 | 4 |   |
|---|---|---|---|---|---|---|---|---|
| 3 | 9 | 8 | 1 | 5 |   | 2 |   | 7 |
|   | 2 | 4 |   | 7 | 8 | 3 |   | 9 |
| 4 | 5 | 1 | 8 | 3 | 6 | 9 |   | 2 |
| 9 | 3 | 2 | 5 | 1 | 7 | 4 | 8 |   |
| 6 |   | 7 | 4 |   | 2 | 1 | 3 |   |
| 8 | 1 |   | 9 |   |   | 7 | 2 | 4 |
| 7 |   | 9 | 2 |   |   |   | 5 |   |
|   |   | 5 |   |   | 3 |   | 9 | 1 |

## Puzzle 45
## Medium

| 1 |   | 5 |   | 2 | 7 |   | 3 |   |
|---|---|---|---|---|---|---|---|---|
|   | 7 |   | 5 | 3 | 9 | 1 |   |   |
| 3 |   | 4 |   | 6 |   |   | 5 | 7 |
|   | 1 | 7 | 2 |   | 5 |   | 8 | 3 |
| 4 |   | 9 | 6 |   | 3 | 2 | 1 | 5 |
| 5 | 3 | 2 | 1 |   | 4 | 6 |   | 9 |
| 2 | 5 | 3 | 9 | 1 | 6 | 7 |   | 8 |
| 7 | 6 | 8 |   | 4 | 2 | 5 | 9 | 1 |
| 9 | 4 | 1 | 7 |   |   |   |   |   |

## Puzzle 46
## Medium

|   |   |   | 1 | 4 | 9 | 7 | 3 | 5 |
|---|---|---|---|---|---|---|---|---|
|   |   | 5 | 8 | 2 | 3 | 6 | 1 | 4 |
|   |   | 4 | 7 | 5 | 6 |   | 8 |   |
|   |   |   | 4 | 9 | 7 | 5 | 6 |   |
| 5 |   |   | 6 | 1 |   | 4 | 7 |   |
| 6 | 4 | 7 | 5 | 3 |   |   | 9 |   |
|   |   |   | 3 |   | 4 | 8 | 5 |   |
|   |   | 3 | 2 | 8 | 5 |   | 4 |   |
| 4 | 5 | 8 | 9 | 6 | 1 | 3 | 2 | 7 |

## Puzzle 47
## Medium

| 7 | 5 | 8 | 2 | 9 | 4 | 3 | 6 | 1 |
|---|---|---|---|---|---|---|---|---|
| 6 | 1 | 2 | 8 | 7 | 3 | 4 |   |   |
| 3 | 4 | 9 |   |   |   | 7 |   | 8 |
| 8 | 6 | 4 | 3 | 2 |   |   | 1 | 7 |
| 9 | 3 |   | 7 | 4 |   |   | 8 | 2 |
| 2 | 7 |   |   |   |   |   |   | 3 |
|   |   |   |   |   |   |   |   | 4 |
| 1 |   | 6 | 4 | 3 | 7 | 2 | 9 | 5 |
| 4 | 9 | 7 | 1 | 5 | 2 | 8 | 3 | 6 |

## Puzzle 48
## Medium

|   |   |   |   | 4 |   | 5 | 2 | 1 |
|---|---|---|---|---|---|---|---|---|
| 4 | 2 | 6 |   |   | 5 |   |   | 9 |
|   | 5 |   |   |   |   | 8 | 4 |   |
| 5 |   |   |   | 1 |   |   |   | 2 |
| 3 | 1 | 4 | 7 | 5 | 2 | 6 | 9 | 8 |
|   | 7 | 2 |   | 6 |   |   |   | 5 |
| 2 | 3 | 1 | 5 | 7 | 8 | 9 | 6 | 4 |
| 7 | 4 | 5 |   |   | 6 | 1 | 8 | 3 |
| 6 | 9 | 8 |   |   | 4 | 2 | 5 | 7 |

## Puzzle 49
### Medium

| | | | | 1 | | 4 | 2 | |
|---|---|---|---|---|---|---|---|---|
| | 4 | | | 2 | | 1 | 6 | |
| 2 | 1 | 3 | 6 | 4 | 7 | 9 | 8 | |
| | 3 | 2 | 4 | 5 | | 7 | | 6 |
| | 9 | | | 3 | 6 | 5 | | 8 |
| | 6 | | | 7 | | 3 | | |
| 3 | 7 | 8 | 1 | 6 | 4 | 2 | 5 | 9 |
| 4 | 2 | 9 | 3 | 8 | 5 | 6 | 7 | 1 |
| | 5 | | 7 | 9 | 2 | 8 | 3 | 4 |

## Puzzle 50
### Medium

| 4 | 1 | 2 | | | 8 | | | 7 |
|---|---|---|---|---|---|---|---|---|
| 9 | | 8 | 5 | 6 | 7 | 2 | 1 | 4 |
| 7 | 5 | 6 | 1 | 4 | 2 | 3 | 8 | |
| 3 | 6 | 9 | 4 | 8 | 5 | 1 | 7 | 2 |
| 8 | 2 | 4 | 7 | 1 | 3 | | | |
| 1 | 7 | 5 | 6 | 2 | 9 | | | |
| 5 | | 7 | | | | | | |
| 2 | | 1 | | 7 | 6 | | | |
| 6 | | 3 | | 5 | | 7 | | |

## Puzzle 51
### Hard

| | | | | 5 | | | 6 | 1 |
|---|---|---|---|---|---|---|---|---|
| | | | | 1 | | | | |
| 1 | | | | 6 | | | | |
| | | 3 | 5 | 4 | 9 | 1 | | 2 |
| | 5 | 1 | 6 | 3 | 2 | | | |
| | 9 | 4 | | | | 6 | 5 | 3 |
| 9 | 7 | 6 | 3 | 2 | 5 | 4 | 1 | 8 |
| 3 | 1 | | | 9 | 6 | 7 | 2 | 5 |
| 5 | 4 | 2 | | | | | | 6 |

## Puzzle 52
### Hard

| 3 | 4 | 7 | 2 | 1 | 6 | 5 | 9 | |
|---|---|---|---|---|---|---|---|---|
| 1 | | 6 | 3 | | | 2 | | |
| | 9 | | | | 5 | 6 | 1 | |
| | 3 | 9 | 6 | | | 4 | 2 | |
| 4 | 7 | 1 | 5 | 2 | | | 3 | 6 |
| 8 | 6 | | | | 4 | 9 | 5 | 7 |
| 6 | | | 9 | 7 | 3 | | | 5 |
| 9 | 8 | | 4 | | | 7 | 6 | |
| 7 | 1 | | | | | | | |

## Puzzle 53
## Hard

|   |   | 3 |   | 5 |   |   | 1 |   |
|---|---|---|---|---|---|---|---|---|
|   |   | 1 |   | 4 |   |   | 3 |   |
|   |   |   | 6 | 1 | 3 |   |   |   |
|   | 3 | 7 |   | 2 |   | 8 | 5 | 1 |
|   |   |   |   | 8 |   | 3 | 2 | 7 |
|   |   |   |   | 3 | 7 | 6 | 9 |   |
|   | 7 | 9 | 8 | 6 |   | 5 | 4 | 3 |
|   |   |   | 3 | 7 | 2 | 1 | 6 | 9 |
| 3 | 1 | 6 | 4 | 9 | 5 | 2 | 7 | 8 |

## Puzzle 54
## Hard

|   |   | 6 |   |   |   |   |   |   |
|---|---|---|---|---|---|---|---|---|
|   |   |   | 8 | 5 |   | 6 |   |   |
|   | 5 |   | 6 |   |   |   |   |   |
| 3 | 2 | 8 | 7 | 9 |   | 4 | 1 |   |
|   |   |   | 3 | 1 |   | 8 |   |   |
|   | 6 |   | 4 | 8 |   | 7 | 3 |   |
| 5 | 8 | 2 |   | 7 |   | 3 |   |   |
| 4 | 7 | 3 | 5 | 6 | 8 | 9 | 2 | 1 |
| 6 | 1 | 9 | 2 | 4 |   |   | 8 | 7 |

## Puzzle 55
### Hard

| | 6 | | | | 2 | | 5 | |
|---|---|---|---|---|---|---|---|---|
| | 2 | | 5 | 6 | 4 | 9 | | |
| | 7 | 5 | | 8 | 3 | 6 | | 2 |
| 5 | 3 | | 2 | 4 | 9 | | 8 | 6 |
| | | | 6 | | 8 | | | 5 |
| | 8 | | | 5 | 1 | | | |
| | 5 | | | | 7 | 4 | 2 | 1 |
| | 4 | | 8 | 1 | 5 | 7 | 6 | |
| 7 | 1 | | 4 | 2 | 6 | 5 | | 8 |

## Puzzle 56
### Hard

| | | | 7 | | | | 4 | |
|---|---|---|---|---|---|---|---|---|
| | | | 3 | | | | 7 | |
| | | | | 9 | | | 8 | |
| 1 | 7 | 2 | 8 | | 9 | 3 | | 4 |
| 3 | 4 | 6 | | | 7 | 8 | | 9 |
| | | | 4 | | 3 | 1 | | 7 |
| | 1 | 4 | 2 | 3 | 5 | | 9 | 8 |
| | | | 9 | 4 | 1 | | 3 | 5 |
| 9 | 3 | 5 | 6 | 7 | 8 | 4 | 1 | 2 |

## Puzzle 57
### Hard

| | | | 5 | 6 | | 7 | 4 | 2 |
|---|---|---|---|---|---|---|---|---|
| 6 | | 7 | | | 4 | 5 | | 9 |
| | 5 | 4 | 9 | | 7 | 6 | | 3 |
| | | 6 | | 1 | 5 | | | 8 |
| 8 | 7 | 2 | 6 | 4 | 3 | 9 | 5 | 1 |
| | | | | 9 | | | | 6 |
| | 3 | | 4 | | 9 | | 6 | 7 |
| 4 | 6 | | | 7 | 2 | | | 5 |
| 7 | | | 1 | | 6 | | | 4 |

## Puzzle 58
### Hard

| | | | | | 9 | | 8 | 5 |
|---|---|---|---|---|---|---|---|---|
| | | | 5 | | 8 | 7 | | 9 |
| | 9 | | | 7 | 2 | | | 4 |
| | | 2 | | | 6 | 5 | 9 | 8 |
| 4 | 5 | 9 | 8 | 2 | 7 | | | |
| | | | 9 | 5 | | 4 | 7 | 2 |
| | | | 2 | | 4 | | 5 | 3 |
| 2 | | 3 | | | 1 | | 4 | |
| 9 | 4 | 5 | 7 | 8 | 3 | 6 | 2 | 1 |

## Puzzle 59
## Hard

| | 3 | | 4 | 8 | | 9 | 6 | 1 |
|---|---|---|---|---|---|---|---|---|
| | 6 | | 9 | 3 | 1 | | | |
| 1 | 9 | | 6 | 7 | | | | 3 |
| | | | | 9 | 8 | 6 | 3 | 4 |
| 3 | 7 | 6 | 1 | 5 | | 2 | 9 | 8 |
| 8 | 4 | 9 | 2 | 6 | 3 | | | |
| | 5 | | 3 | | | 1 | 8 | |
| 6 | | | 8 | 4 | | | | |
| 9 | 8 | | 5 | 1 | | | 4 | |

## Puzzle 60
## Hard

| 3 | 9 | 4 | 5 | 6 | 2 | 8 | 7 | 1 |
|---|---|---|---|---|---|---|---|---|
| 2 | | | 8 | 9 | 7 | 5 | 4 | 3 |
| 8 | 7 | 5 | 4 | 3 | 1 | 6 | 9 | 2 |
| | | | 6 | | 5 | | 3 | 8 |
| 6 | 3 | 8 | | 1 | 9 | | | 4 |
| 5 | | 7 | 3 | | 8 | 1 | 6 | 9 |
| | 8 | | | | 6 | | 1 | 5 |
| | | | | | | | | |
| | | | | | | | | |

## Puzzle 61
### Hard

| | 8 | 4 | 6 | 5 | 2 | | | 1 |
|---|---|---|---|---|---|---|---|---|
| | 5 | | | 8 | 1 | | | 2 |
| 6 | | | 9 | 7 | 3 | 5 | 8 | 4 |
| | 6 | | | 2 | 4 | 1 | | 9 |
| 5 | | | | 9 | 6 | 2 | 4 | |
| 2 | 4 | 9 | | 1 | | | | |
| 1 | | | | 4 | | | | 6 |
| 4 | | 6 | 1 | 3 | 7 | | | |
| 8 | 7 | 5 | 2 | 6 | 9 | 4 | | |

## Puzzle 62
### Hard

| 9 | 3 | 5 | | | 8 | | 6 | |
|---|---|---|---|---|---|---|---|---|
| | 7 | 8 | | | | 5 | 3 | 9 |
| | 2 | 6 | | 9 | 5 | 7 | 8 | |
| 2 | | 9 | 1 | | | 8 | 7 | 5 |
| 7 | 5 | 3 | | 8 | | 1 | 9 | |
| 8 | 1 | | | 5 | 7 | | 2 | 6 |
| | 9 | | | 7 | 1 | | 4 | |
| 5 | 8 | | 4 | | 9 | | | |
| | | 1 | | | | 9 | 5 | 7 |

## Puzzle 63
## Hard

| | | | | | | | 1 | 3 |
|---|---|---|---|---|---|---|---|---|
| 5 | 9 | 8 | 6 | 3 | 1 | 7 | 2 | |
| 1 | 4 | 3 | | 2 | 5 | 8 | | 9 |
| | | | | | | | | 8 |
| | 5 | | | 6 | | 4 | 3 | |
| | | | | | | | | |
| 4 | 8 | 5 | 2 | 1 | 3 | 9 | 7 | 6 |
| | | 1 | 8 | 7 | 6 | 2 | 4 | 5 |
| 7 | 6 | 2 | | | | 3 | 8 | 1 |

## Puzzle 64
## Hard

| | | | 1 | 2 | 8 | 7 | 6 | 4 |
|---|---|---|---|---|---|---|---|---|
| | | 6 | 9 | 7 | 4 | 2 | | 3 |
| | 4 | | 3 | 6 | 5 | 8 | 1 | 9 |
| | | 5 | 4 | 3 | | | | 6 |
| 3 | 6 | | | 9 | | | 4 | |
| | | | 6 | 8 | | 5 | 3 | |
| | 7 | | 2 | | 6 | 3 | | |
| | | | 8 | 1 | 9 | | 7 | 5 |
| | | | 7 | | 3 | | | |

## Puzzle 65
### Hard

| 9 |   | 6 |   | 3 | 8 | 7 | 4 | 2 |
|---|---|---|---|---|---|---|---|---|
| 7 |   | 3 | 2 | 4 | 6 |   |   |   |
| 8 | 4 | 2 | 7 | 9 |   |   | 6 |   |
| 5 |   | 1 |   | 2 |   |   |   |   |
| 3 |   | 7 | 4 | 5 |   | 6 | 2 |   |
| 2 |   | 4 |   | 8 | 7 |   |   | 5 |
|   |   | 9 |   |   |   |   |   |   |
|   | 3 | 8 | 9 |   | 4 | 2 | 5 |   |
| 4 |   | 5 | 8 | 1 | 2 | 9 |   | 6 |

## Puzzle 66
### Hard

|   | 8 |   |   | 2 | 6 |   | 9 | 7 |
|---|---|---|---|---|---|---|---|---|
| 2 | 7 | 6 |   | 8 | 9 | 3 |   |   |
|   |   | 4 | 1 | 3 | 7 | 8 |   | 6 |
| 6 |   |   | 3 | 1 | 8 | 5 | 7 |   |
|   |   |   |   | 9 |   |   |   | 4 |
|   |   | 2 | 6 | 4 | 5 | 1 | 8 | 9 |
| 9 |   | 7 | 8 | 6 | 3 |   |   |   |
| 4 |   |   | 9 |   | 1 |   | 6 | 8 |
| 8 |   |   |   |   |   |   | 1 | 3 |

## Puzzle 67
### Hard

| 1 | 8 | 4 | 7 | 2 | 9 |   | 6 |   |
|---|---|---|---|---|---|---|---|---|
| 7 | 3 | 5 |   |   |   | 2 | 9 | 1 |
| 9 | 6 | 2 |   |   |   | 8 | 7 | 4 |
|   |   | 6 |   |   |   |   |   |   |
|   |   | 8 |   |   |   |   |   |   |
|   | 7 | 1 |   |   |   |   | 5 | 2 |
| 6 | 5 | 7 |   |   | 8 | 4 | 2 | 9 |
| 8 | 1 | 9 | 5 | 4 | 2 | 7 | 3 | 6 |
| 2 | 4 | 3 |   |   |   |   |   |   |

## Puzzle 68
### Hard

| 8 |   | 1 |   |   | 2 | 7 |   |   |
|---|---|---|---|---|---|---|---|---|
| 9 |   |   |   |   |   |   |   |   |
| 2 |   |   |   |   |   | 8 |   | 3 |
| 3 | 8 | 2 |   |   |   |   |   |   |
| 1 | 5 | 7 |   |   | 3 |   |   |   |
| 6 | 4 | 9 | 2 |   | 5 | 3 |   | 8 |
| 7 | 2 | 6 |   |   |   | 1 | 3 | 5 |
| 4 | 1 | 8 |   |   | 7 | 2 | 6 | 9 |
| 5 | 9 | 3 | 6 | 2 | 1 | 4 | 8 | 7 |

## Puzzle 69
## Hard

| 5 | 3 | 6 |   |   | 1 | 9 |   |   |
|---|---|---|---|---|---|---|---|---|
|   | 4 | 1 | 7 |   |   |   | 6 | 5 |
|   | 7 |   |   |   | 6 |   |   |   |
|   | 6 |   |   |   | 9 |   | 5 |   |
|   | 9 | 7 | 2 |   | 5 | 6 |   |   |
| 3 | 1 | 5 | 6 | 7 | 8 |   | 9 |   |
| 7 | 8 | 9 |   |   | 4 | 5 | 2 |   |
| 1 |   | 3 | 5 | 6 | 7 | 8 | 4 | 9 |
| 6 | 5 | 4 |   |   |   |   |   |   |

## Puzzle 70
## Hard

| 5 | 4 |   | 7 | 9 |   |   | 8 | 3 |
|---|---|---|---|---|---|---|---|---|
|   |   |   | 3 | 5 |   |   |   | 6 |
|   | 3 | 7 | 4 | 8 | 1 | 9 | 2 | 5 |
| 7 | 6 |   | 1 |   | 9 | 8 |   |   |
| 1 |   |   | 2 |   |   |   |   |   |
|   |   |   | 6 |   |   |   | 1 |   |
| 4 | 7 | 6 |   | 2 |   | 5 | 9 | 1 |
| 2 |   |   |   | 6 |   | 7 | 3 | 8 |
| 3 |   | 8 | 9 | 1 | 7 | 2 | 6 | 4 |

## Puzzle 71
### Hard

| | | | 5 | | | | 1 | |
|---|---|---|---|---|---|---|---|---|
| | 5 | 2 | | | | 9 | | |
| | 8 | 7 | | 9 | | | | |
| | | 5 | | | 9 | | | |
| 7 | | | | 8 | | | | |
| | 4 | 8 | 1 | 5 | | 3 | 9 | |
| 9 | 7 | 1 | 2 | 4 | 5 | 6 | 8 | 3 |
| 5 | 2 | 3 | 8 | 1 | 6 | 7 | 4 | 9 |
| 8 | 6 | 4 | 9 | 7 | 3 | 2 | 5 | 1 |

## Puzzle 72
### Hard

| | | | 4 | | 8 | 1 | 6 | |
|---|---|---|---|---|---|---|---|---|
| | | | | | 7 | 4 | 8 | |
| 9 | 8 | 4 | | | | 2 | 7 | |
| 8 | 1 | 2 | 5 | 6 | 3 | 9 | 4 | 7 |
| 5 | 7 | 6 | 9 | 4 | 2 | 8 | 3 | 1 |
| 3 | 4 | 9 | 7 | 8 | 1 | 6 | 5 | 2 |
| | | | | 7 | 9 | | 2 | |
| | 9 | | | | | | | |
| | | | 8 | | 5 | | 9 | |

## Puzzle 73
### Hard

| | | | 8 | 6 | 1 | | 7 | 4 |
|---|---|---|---|---|---|---|---|---|
| 7 | | | 4 | | 5 | | | 3 |
| | | | 7 | | | 5 | | 8 |
| | | | | 4 | | | 9 | 7 |
| | 4 | | 2 | | 7 | | 3 | 5 |
| | 7 | | | | | | 4 | 2 |
| | 2 | | 3 | | 6 | 7 | 5 | 1 |
| 3 | 6 | 7 | 1 | 5 | 2 | 4 | 8 | 9 |
| 5 | | | 9 | 7 | 4 | 3 | 2 | 6 |

## Puzzle 74
### Hard

| | | | 2 | 7 | 1 | 3 | 4 | |
|---|---|---|---|---|---|---|---|---|
| | | | 6 | 4 | 5 | | | 8 |
| | | | 3 | 9 | 8 | | | |
| | | | 7 | 6 | | | 8 | 3 |
| 8 | 6 | | | 5 | | | | |
| | | 7 | 1 | 8 | | | | |
| | | | | 2 | | | 3 | |
| 7 | 4 | 2 | 8 | 3 | 6 | 9 | 5 | 1 |
| 9 | 3 | 6 | 5 | 1 | 7 | 8 | 2 | 4 |

## Puzzle 75
### Hard

| 2 | 6 | 7 |   |   |   | 8 | 1 |   |
|---|---|---|---|---|---|---|---|---|
| 4 | 8 | 1 |   |   |   | 2 | 3 |   |
| 9 | 5 | 3 | 1 | 2 | 8 | 4 | 7 | 6 |
|   | 1 |   |   | 4 |   |   | 9 |   |
| 3 | 7 |   |   |   |   | 5 | 6 | 8 |
| 6 | 2 | 9 | 8 |   |   |   | 4 |   |
| 7 | 3 | 8 |   |   |   |   | 5 |   |
| 1 |   | 6 |   | 8 |   |   | 2 | 3 |
| 5 |   | 2 |   |   |   |   | 8 |   |

## Puzzle 76
### Fiendish

|   |   | 4 | 1 | 2 |   | 7 | 8 | 5 |
|---|---|---|---|---|---|---|---|---|
| 1 |   |   | 4 | 8 | 5 |   |   |   |
| 5 | 2 | 8 | 3 | 9 | 7 |   | 6 |   |
| 8 | 4 | 1 |   |   |   | 9 | 7 | 6 |
|   |   |   | 8 | 1 |   |   |   |   |
|   |   |   | 7 | 6 |   |   | 1 | 8 |
|   | 1 | 2 | 9 |   |   |   | 3 |   |
| 4 |   |   |   |   |   |   |   |   |
|   |   |   |   |   |   |   |   |   |

## Puzzle 77
## Fiendish

| | | | | | | | 3 | |
|---|---|---|---|---|---|---|---|---|
| | 3 | | | | | 1 | 5 | |
| 5 | | | | 9 | | 4 | 8 | 7 |
| | | | 9 | | 7 | | 2 | |
| | 9 | | 5 | | | | 6 | |
| 3 | | | 8 | 1 | 6 | 7 | 9 | 5 |
| | 8 | | 1 | 6 | | 2 | 7 | 3 |
| | 6 | | | | | 5 | 1 | 8 |
| | | | | | | 9 | 4 | 6 |

## Puzzle 78
## Fiendish

| 7 | 1 | 4 | 5 | | 3 | 9 | | |
|---|---|---|---|---|---|---|---|---|
| | | 2 | | | | 5 | | |
| 9 | 5 | 8 | | | 7 | | | 3 |
| 5 | | 1 | | | 6 | | | 9 |
| | | 7 | | | | | | 5 |
| | | 9 | | | 5 | | 3 | |
| 1 | 9 | 3 | | | | | 5 | |
| 2 | 7 | 5 | 3 | | | | 6 | |
| | | | | | 1 | 3 | 9 | |

## Puzzle 79
### Fiendish

|   | 4 | 2 |   |   | 3 | 5 | 6 |   |
|---|---|---|---|---|---|---|---|---|
|   |   |   | 9 |   | 6 |   |   | 4 |
|   | 8 | 6 | 2 | 5 | 4 |   |   |   |
| 4 |   | 3 |   |   |   |   |   |   |
|   |   |   |   |   |   |   |   |   |
|   | 9 |   |   | 2 |   | 6 | 4 |   |
| 2 | 6 | 7 |   | 4 | 8 | 1 |   |   |
| 5 | 3 | 4 | 6 |   |   | 8 | 7 | 2 |

## Puzzle 80
### Fiendish

|   | 4 |   |   | 1 | 9 |   |   |   |
|---|---|---|---|---|---|---|---|---|
|   | 2 |   |   |   |   | 9 |   |   |
|   | 8 |   |   |   | 5 |   |   | 1 |
|   |   |   |   |   |   | 5 | 9 | 4 |
| 5 | 3 | 4 | 6 | 9 | 1 | 8 | 7 | 2 |
| 2 | 9 | 7 | 8 | 5 | 4 |   |   |   |
| 4 |   | 3 | 1 |   |   | 7 |   | 9 |
|   |   | 2 |   |   |   |   |   |   |
|   |   |   |   |   |   |   |   |   |

## Puzzle 81
## Fiendish

| | | | 6 | | | | 9 | 1 |
|---|---|---|---|---|---|---|---|---|
| 9 | | 4 | | | | 2 | 8 | 6 |
| | | | 9 | | | | | |
| | 9 | | 4 | 8 | 1 | | | |
| | 4 | 8 | 7 | | | 9 | 1 | |
| 1 | 6 | 3 | 2 | 5 | 9 | 8 | 7 | 4 |
| | | | | 9 | | | | 8 |
| | | | | | | | | 9 |
| | 2 | 9 | | | 4 | | 3 | 7 |

## Puzzle 82
## Fiendish

| | 1 | | | | | 5 | | |
|---|---|---|---|---|---|---|---|---|
| 7 | | | | | | 4 | | |
| | 5 | | | 9 | 6 | 2 | 3 | |
| | 2 | | 6 | | | 9 | | 3 |
| | | | | | 2 | 7 | | 5 |
| | 3 | | | | 9 | 8 | 2 | 6 |
| 3 | 7 | 6 | 8 | 2 | 5 | 1 | 9 | 4 |
| | | 1 | 9 | | | | | 2 |
| | | | | | | | | |

## Puzzle 83
## Fiendish

|   |   |   |   |   |   |   |   |   |
|---|---|---|---|---|---|---|---|---|
|   |   |   |   |   |   |   |   |   |
|   |   |   |   |   |   |   |   |   |
| 3 | 1 | 8 |   | 9 |   |   | 5 |   |
|   |   | 1 |   |   |   |   |   |   |
|   | 6 |   |   |   | 1 | 5 |   |   |
| 7 | 9 | 4 | 5 | 8 | 2 | 3 | 6 | 1 |
|   |   |   |   |   |   | 1 | 4 | 5 |
| 1 |   |   |   |   | 5 |   |   |   |
| 9 | 4 | 5 | 1 | 7 | 8 | 2 | 3 | 6 |

## Puzzle 84
## Fiendish

|   |   |   |   |   |   |   |   |   |
|---|---|---|---|---|---|---|---|---|
| 5 | 1 | 9 |   |   |   | 6 | 8 | 4 |
| 7 | 8 | 6 |   |   |   |   | 5 |   |
| 3 | 4 |   | 6 |   | 8 |   | 1 |   |
|   |   | 8 |   |   | 5 |   |   |   |
|   |   |   | 1 |   | 8 |   |   |   |
|   | 3 |   |   |   |   | 8 |   |   |
| 6 |   |   | 8 | 3 |   |   |   |   |
|   |   | 3 | 5 |   | 7 |   |   | 8 |
| 8 | 5 | 4 | 2 |   |   |   |   |   |

## Puzzle 85
## Fiendish

| 1 |   |   |   |   |   |   |   |   |
|---|---|---|---|---|---|---|---|---|
|   | 8 |   | 3 | 1 |   |   | 4 |   |
| 9 | 2 | 4 | 7 | 3 | 1 | 6 | 5 |   |
|   |   |   | 4 | 9 | 8 |   | 7 |   |
| 8 |   |   | 5 |   | 6 | 4 | 9 |   |
|   |   | 1 |   |   |   |   | 6 | 9 |
| 4 | 9 | 8 | 1 | 6 | 3 | 5 | 2 | 7 |
| 6 |   |   |   |   |   |   |   | 4 |

## Puzzle 86
## Fiendish

|   |   | 2 |   | 7 |   |   | 4 |   |
|---|---|---|---|---|---|---|---|---|
|   |   |   |   |   | 4 |   |   |   |
| 5 | 8 | 4 |   |   |   | 9 | 7 |   |
|   |   |   | 3 | 5 |   | 2 | 9 | 4 |
| 3 | 5 | 9 |   |   |   |   |   |   |
| 2 | 4 |   |   | 1 | 9 | 7 | 5 | 3 |
|   |   | 3 |   |   |   | 4 | 2 | 5 |
| 4 | 6 |   | 2 |   |   |   |   | 9 |
|   | 2 |   |   |   |   | 6 | 8 | 7 |

## Puzzle 87
### Fiendish

| | | | | | 1 | | | |
|---|---|---|---|---|---|---|---|---|
| | | | 8 | 3 | | 1 | | |
| 6 | 1 | | | | 7 | 3 | | |
| 3 | 6 | 2 | 9 | 1 | 5 | | | |
| | 4 | | 3 | 8 | 6 | 5 | 2 | 1 |
| | 5 | 1 | 2 | 7 | 4 | 6 | 3 | |
| | 2 | | | 5 | | 9 | | |
| | | | | | 3 | 2 | | 5 |
| | 3 | | 1 | | | | | |

## Puzzle 88
### Fiendish

| 3 | | | 9 | 5 | 6 | | 2 | |
|---|---|---|---|---|---|---|---|---|
| | | | | | | | | 3 |
| | | 2 | 3 | 4 | 1 | | | |
| 9 | 5 | 3 | | 6 | 7 | | 8 | |
| 4 | | | | 3 | | | | 9 |
| | 2 | | | 9 | 4 | | 3 | |
| | | | | | | 3 | | |
| | 3 | 5 | | | | | | 8 |
| | | 8 | 6 | 7 | 3 | 5 | | |

## Puzzle 89
### Fiendish

| | | | | | 1 | | | |
|---|---|---|---|---|---|---|---|---|
| | | 1 | 8 | 2 | 6 | 4 | 9 | 7 |
| | | 2 | | 5 | 7 | | | |
| | | | | | | | | 9 |
| | 4 | | 6 | 9 | 2 | | 5 | 8 |
| 8 | 9 | | 7 | 1 | 3 | 2 | 6 | 4 |
| | | 4 | 1 | 6 | 9 | 3 | | |
| | | 9 | 2 | | | | | |
| | | | | | | 9 | | |

## Puzzle 90
### Fiendish

| 7 | 8 | 6 | | | | | 1 | |
|---|---|---|---|---|---|---|---|---|
| 5 | 1 | | 7 | | | | 2 | |
| 9 | 2 | | | | | | 7 | |
| | 9 | 7 | | | 4 | | 8 | |
| | 4 | | | 9 | | | 3 | 2 |
| | | 2 | | | | | | |
| 2 | 7 | 9 | 8 | | | 3 | 5 | 1 |
| | | 1 | | | | | | |
| 4 | | 8 | | | | 2 | 6 | 7 |

## Puzzle 91
### Fiendish

| 1 | 3 |   |   |   | 4 | 5 |   |   |
|---|---|---|---|---|---|---|---|---|
| 2 |   | 4 |   | 3 |   |   |   |   |
|   |   | 5 |   | 9 | 8 | 2 | 1 | 6 |
|   |   |   | 1 |   | 5 |   |   |   |
| 8 |   | 1 | 4 | 2 | 6 | 3 | 5 | 7 |
|   | 8 | 7 | 6 |   | 3 |   |   |   |
| 6 | 1 | 3 | 9 | 4 | 2 |   |   | 5 |
|   |   |   |   |   | 7 |   |   |   |
|   |   |   |   |   |   |   |   |   |

## Puzzle 92
### Fiendish

| 2 | 7 |   |   | 9 | 1 | 3 |   | 5 |
|---|---|---|---|---|---|---|---|---|
| 6 | 5 |   |   | 2 |   | 1 |   |   |
| 1 |   |   |   |   | 5 |   |   |   |
| 9 |   |   |   |   |   |   | 2 |   |
| 5 |   | 2 | 1 | 8 | 9 |   |   | 7 |
|   | 3 | 1 |   |   | 2 |   |   | 6 |
|   |   |   | 9 |   |   |   | 1 |   |
|   | 1 |   |   |   |   |   | 9 |   |
|   | 9 | 6 | 2 | 1 |   |   |   | 8 |

## Puzzle 93
### Fiendish

| 4 |   |   | 9 | 1 | 5 |   | 2 | 3 |
|---|---|---|---|---|---|---|---|---|
|   | 3 |   |   | 7 | 8 |   |   |   |
|   | 1 |   |   | 3 | 6 |   |   |   |
| 8 |   |   | 3 | 9 | 7 | 2 | 6 |   |
| 6 | 2 | 3 | 1 |   | 4 |   |   |   |
|   |   |   | 5 | 6 | 2 | 3 |   | 8 |
|   |   |   | 7 |   |   |   |   |   |
| 3 | 7 |   | 6 |   |   |   |   | 2 |
|   |   |   | 8 |   |   |   |   |   |

## Puzzle 94
### Fiendish

| 7 |   |   | 1 |   | 3 |   |   | 8 |
|---|---|---|---|---|---|---|---|---|
|   | 8 |   |   |   |   | 7 |   | 1 |
| 6 |   | 1 | 8 | 4 | 7 | 3 |   |   |
|   |   |   |   | 1 | 4 | 8 | 2 |   |
|   | 1 |   |   |   |   | 5 | 7 |   |
| 2 | 3 |   | 7 | 6 |   | 1 |   |   |
| 1 |   |   |   |   | 9 | 4 | 3 | 7 |
|   | 4 |   |   |   |   |   |   |   |
|   |   |   | 4 |   |   |   |   |   |

## Puzzle 95
## Fiendish

| 4 |   |   | 6 | 2 |   |   | 3 |   |
|---|---|---|---|---|---|---|---|---|
| 6 |   | 2 |   | 3 |   |   |   |   |
|   |   |   |   | 4 | 9 |   | 6 |   |
| 8 | 4 | 7 |   | 6 |   |   | 2 |   |
| 1 | 2 | 6 |   |   | 4 |   |   |   |
| 9 |   |   | 1 | 7 | 2 | 6 |   | 4 |
|   |   |   | 4 |   |   |   |   | 6 |
|   |   |   |   |   | 6 | 7 | 9 |   |
|   | 6 |   |   |   |   | 4 |   |   |

## Puzzle 96
## Fiendish

| 1 |   |   | 8 | 4 | 7 |   | 9 |   |
|---|---|---|---|---|---|---|---|---|
|   |   |   | 5 | 3 | 9 |   |   |   |
|   |   |   | 1 | 2 | 6 | 3 |   | 4 |
|   | 1 |   |   | 5 | 8 |   | 4 |   |
|   |   |   | 2 | 1 | 3 | 7 |   |   |
| 2 |   |   | 7 | 9 | 4 |   |   |   |
|   | 4 | 1 |   | 7 | 2 |   | 6 |   |
| 6 |   |   |   | 8 |   |   |   |   |
|   |   |   |   | 6 | 1 |   |   |   |

## Puzzle 97
### Fiendish

| 4 | 7 |   |   | 8 |   | 9 |   |   |
|---|---|---|---|---|---|---|---|---|
| 3 | 5 |   |   | 9 |   | 8 | 1 | 7 |
|   | 1 |   | 7 | 2 |   | 5 | 4 |   |
| 6 |   | 3 |   | 4 |   |   | 5 |   |
| 2 |   | 5 | 3 | 6 |   |   |   |   |
|   | 8 |   | 2 | 5 | 9 |   | 6 | 3 |
|   | 3 |   |   |   | 2 |   | 8 | 4 |
|   |   | 4 |   |   |   |   |   |   |
|   |   |   |   |   |   |   |   |   |

## Puzzle 98
### Fiendish

| 9 |   | 5 | 2 |   | 8 |   |   |   |
|---|---|---|---|---|---|---|---|---|
|   |   | 2 |   |   |   | 9 | 8 |   |
| 1 |   | 8 | 9 |   | 7 |   |   | 3 |
| 5 | 1 |   |   | 2 |   |   |   | 8 |
|   |   |   | 8 |   | 1 | 6 | 5 | 2 |
| 8 | 2 |   |   | 7 |   | 1 |   | 9 |
| 7 |   | 1 | 6 | 8 | 2 |   |   | 4 |
|   | 8 |   |   |   |   |   |   |   |
| 2 |   |   |   | 1 |   | 8 |   |   |

## Puzzle 99
## Fiendish

| 4 |   | 2 |   |   |   | 6 |   |   |
|---|---|---|---|---|---|---|---|---|
| 1 | 5 | 8 |   | 6 | 4 | 9 | 3 |   |
| 3 |   | 6 |   | 8 |   |   |   | 4 |
| 6 |   | 1 |   |   |   | 5 | 7 |   |
| 5 | 2 | 4 |   |   |   | 3 |   |   |
| 9 |   | 7 |   |   |   | 4 | 1 |   |
|   | 1 |   |   |   |   |   | 4 |   |
| 7 | 4 | 9 |   |   | 3 |   | 6 | 5 |
|   | 6 |   |   |   |   |   |   |   |

## Puzzle 100
## Fiendish

|   |   |   | 8 |   |   |   |   | 9 |
|---|---|---|---|---|---|---|---|---|
|   |   |   | 9 |   |   |   |   |   |
|   |   | 9 | 7 |   |   | 5 |   | 8 |
|   | 3 |   |   |   |   |   |   |   |
| 9 |   | 2 |   | 4 |   | 3 |   | 7 |
|   |   |   |   |   |   |   |   |   |
|   | 9 |   | 4 | 2 | 7 | 1 |   |   |
| 7 | 2 | 1 |   |   |   | 6 | 9 | 4 |
| 3 | 4 | 5 | 1 | 6 | 9 | 8 | 7 | 2 |

# Solutions

## Puzzle 1

| 7 | 4 | 2 | 8 | 3 | 9 | 6 | 5 | 1 |
|---|---|---|---|---|---|---|---|---|
| 1 | 9 | 5 | 4 | 6 | 2 | 8 | 7 | 3 |
| 8 | 6 | 3 | 1 | 5 | 7 | 2 | 4 | 9 |
| 5 | 2 | 4 | 3 | 7 | 8 | 9 | 1 | 6 |
| 3 | 8 | 1 | 5 | 9 | 6 | 7 | 2 | 4 |
| 6 | 7 | 9 | 2 | 1 | 4 | 5 | 3 | 8 |
| 2 | 5 | 8 | 9 | 4 | 3 | 1 | 6 | 7 |
| 4 | 1 | 6 | 7 | 8 | 5 | 3 | 9 | 2 |
| 9 | 3 | 7 | 6 | 2 | 1 | 4 | 8 | 5 |

## Puzzle 2

| 8 | 9 | 5 | 1 | 7 | 3 | 6 | 4 | 2 |
|---|---|---|---|---|---|---|---|---|
| 7 | 4 | 3 | 6 | 8 | 2 | 1 | 9 | 5 |
| 6 | 2 | 1 | 4 | 5 | 9 | 3 | 8 | 7 |
| 4 | 6 | 2 | 9 | 3 | 5 | 7 | 1 | 8 |
| 5 | 1 | 7 | 8 | 4 | 6 | 9 | 2 | 3 |
| 3 | 8 | 9 | 2 | 1 | 7 | 5 | 6 | 4 |
| 1 | 3 | 4 | 5 | 6 | 8 | 2 | 7 | 9 |
| 2 | 7 | 6 | 3 | 9 | 4 | 8 | 5 | 1 |
| 9 | 5 | 8 | 7 | 2 | 1 | 4 | 3 | 6 |

## Puzzle 3

| 4 | 5 | 9 | 6 | 3 | 1 | 8 | 2 | 7 |
|---|---|---|---|---|---|---|---|---|
| 7 | 6 | 8 | 9 | 5 | 2 | 4 | 1 | 3 |
| 3 | 1 | 2 | 8 | 7 | 4 | 5 | 6 | 9 |
| 1 | 3 | 4 | 7 | 2 | 5 | 6 | 9 | 8 |
| 2 | 9 | 7 | 1 | 6 | 8 | 3 | 5 | 4 |
| 5 | 8 | 6 | 3 | 4 | 9 | 2 | 7 | 1 |
| 8 | 7 | 5 | 2 | 1 | 3 | 9 | 4 | 6 |
| 9 | 4 | 1 | 5 | 8 | 6 | 7 | 3 | 2 |
| 6 | 2 | 3 | 4 | 9 | 7 | 1 | 8 | 5 |

## Puzzle 4

| 2 | 3 | 5 | 1 | 8 | 6 | 4 | 9 | 7 |
|---|---|---|---|---|---|---|---|---|
| 4 | 8 | 7 | 2 | 9 | 5 | 6 | 1 | 3 |
| 1 | 6 | 9 | 4 | 3 | 7 | 8 | 5 | 2 |
| 8 | 7 | 6 | 9 | 4 | 1 | 3 | 2 | 5 |
| 5 | 9 | 4 | 3 | 7 | 2 | 1 | 6 | 8 |
| 3 | 2 | 1 | 6 | 5 | 8 | 9 | 7 | 4 |
| 7 | 4 | 3 | 5 | 1 | 9 | 2 | 8 | 6 |
| 6 | 1 | 8 | 7 | 2 | 4 | 5 | 3 | 9 |
| 9 | 5 | 2 | 8 | 6 | 3 | 7 | 4 | 1 |

## Puzzle 5

| 9 | 8 | 1 | 5 | 4 | 7 | 2 | 3 | 6 |
|---|---|---|---|---|---|---|---|---|
| 2 | 5 | 3 | 6 | 8 | 9 | 4 | 7 | 1 |
| 4 | 6 | 7 | 3 | 2 | 1 | 8 | 5 | 9 |
| 7 | 4 | 5 | 8 | 9 | 2 | 6 | 1 | 3 |
| 6 | 1 | 2 | 7 | 3 | 5 | 9 | 8 | 4 |
| 8 | 3 | 9 | 4 | 1 | 6 | 5 | 2 | 7 |
| 1 | 2 | 8 | 9 | 6 | 3 | 7 | 4 | 5 |
| 3 | 7 | 6 | 2 | 5 | 4 | 1 | 9 | 8 |
| 5 | 9 | 4 | 1 | 7 | 8 | 3 | 6 | 2 |

## Puzzle 6

| 3 | 8 | 9 | 2 | 7 | 6 | 1 | 4 | 5 |
|---|---|---|---|---|---|---|---|---|
| 4 | 2 | 6 | 3 | 5 | 1 | 9 | 8 | 7 |
| 5 | 1 | 7 | 4 | 8 | 9 | 2 | 6 | 3 |
| 6 | 4 | 1 | 8 | 3 | 7 | 5 | 2 | 9 |
| 2 | 5 | 3 | 9 | 1 | 4 | 8 | 7 | 6 |
| 7 | 9 | 8 | 5 | 6 | 2 | 4 | 3 | 1 |
| 1 | 3 | 4 | 7 | 2 | 5 | 6 | 9 | 8 |
| 9 | 7 | 5 | 6 | 4 | 8 | 3 | 1 | 2 |
| 8 | 6 | 2 | 1 | 9 | 3 | 7 | 5 | 4 |

## Puzzle 7

| 7 | 8 | 9 | 5 | 2 | 4 | 1 | 3 | 6 |
|---|---|---|---|---|---|---|---|---|
| 3 | 1 | 2 | 9 | 6 | 8 | 4 | 7 | 5 |
| 6 | 4 | 5 | 7 | 3 | 1 | 8 | 2 | 9 |
| 5 | 7 | 3 | 6 | 8 | 9 | 2 | 1 | 4 |
| 1 | 2 | 6 | 3 | 4 | 7 | 9 | 5 | 8 |
| 8 | 9 | 4 | 2 | 1 | 5 | 3 | 6 | 7 |
| 2 | 5 | 1 | 4 | 9 | 6 | 7 | 8 | 3 |
| 4 | 3 | 7 | 8 | 5 | 2 | 6 | 9 | 1 |
| 9 | 6 | 8 | 1 | 7 | 3 | 5 | 4 | 2 |

## Puzzle 8

| 3 | 2 | 5 | 9 | 7 | 6 | 8 | 1 | 4 |
|---|---|---|---|---|---|---|---|---|
| 1 | 8 | 9 | 5 | 4 | 2 | 7 | 3 | 6 |
| 7 | 6 | 4 | 1 | 3 | 8 | 9 | 2 | 5 |
| 6 | 4 | 7 | 2 | 9 | 3 | 5 | 8 | 1 |
| 8 | 5 | 2 | 7 | 6 | 1 | 3 | 4 | 9 |
| 9 | 3 | 1 | 4 | 8 | 5 | 6 | 7 | 2 |
| 4 | 1 | 8 | 3 | 5 | 9 | 2 | 6 | 7 |
| 5 | 7 | 6 | 8 | 2 | 4 | 1 | 9 | 3 |
| 2 | 9 | 3 | 6 | 1 | 7 | 4 | 5 | 8 |

## Puzzle 9

| 8 | 4 | 6 | 1 | 7 | 3 | 2 | 9 | 5 |
|---|---|---|---|---|---|---|---|---|
| 3 | 1 | 9 | 6 | 2 | 5 | 7 | 4 | 8 |
| 5 | 2 | 7 | 9 | 4 | 8 | 6 | 1 | 3 |
| 4 | 9 | 3 | 2 | 6 | 1 | 8 | 5 | 7 |
| 1 | 5 | 2 | 4 | 8 | 7 | 3 | 6 | 9 |
| 7 | 6 | 8 | 5 | 3 | 9 | 4 | 2 | 1 |
| 9 | 3 | 4 | 8 | 1 | 6 | 5 | 7 | 2 |
| 6 | 7 | 5 | 3 | 9 | 2 | 1 | 8 | 4 |
| 2 | 8 | 1 | 7 | 5 | 4 | 9 | 3 | 6 |

## Puzzle 10

| 2 | 3 | 5 | 9 | 6 | 1 | 7 | 8 | 4 |
|---|---|---|---|---|---|---|---|---|
| 8 | 7 | 9 | 3 | 4 | 2 | 6 | 5 | 1 |
| 1 | 4 | 6 | 5 | 7 | 8 | 9 | 3 | 2 |
| 6 | 2 | 4 | 8 | 9 | 5 | 1 | 7 | 3 |
| 3 | 5 | 1 | 7 | 2 | 4 | 8 | 6 | 9 |
| 9 | 8 | 7 | 1 | 3 | 6 | 4 | 2 | 5 |
| 7 | 1 | 2 | 4 | 8 | 3 | 5 | 9 | 6 |
| 4 | 9 | 3 | 6 | 5 | 7 | 2 | 1 | 8 |
| 5 | 6 | 8 | 2 | 1 | 9 | 3 | 4 | 7 |

## Puzzle 11

| 6 | 8 | 7 | 4 | 5 | 9 | 2 | 3 | 1 |
|---|---|---|---|---|---|---|---|---|
| 3 | 9 | 4 | 2 | 1 | 6 | 8 | 5 | 7 |
| 2 | 1 | 5 | 7 | 8 | 3 | 9 | 4 | 6 |
| 8 | 7 | 3 | 1 | 2 | 5 | 6 | 9 | 4 |
| 9 | 4 | 2 | 6 | 3 | 8 | 1 | 7 | 5 |
| 5 | 6 | 1 | 9 | 4 | 7 | 3 | 2 | 8 |
| 1 | 2 | 9 | 8 | 7 | 4 | 5 | 6 | 3 |
| 4 | 3 | 8 | 5 | 6 | 2 | 7 | 1 | 9 |
| 7 | 5 | 6 | 3 | 9 | 1 | 4 | 8 | 2 |

## Puzzle 12

| 5 | 6 | 8 | 9 | 3 | 2 | 4 | 7 | 1 |
|---|---|---|---|---|---|---|---|---|
| 4 | 3 | 9 | 7 | 1 | 5 | 6 | 2 | 8 |
| 7 | 1 | 2 | 8 | 6 | 4 | 5 | 3 | 9 |
| 2 | 8 | 6 | 5 | 7 | 3 | 1 | 9 | 4 |
| 9 | 5 | 1 | 2 | 4 | 8 | 3 | 6 | 7 |
| 3 | 7 | 4 | 1 | 9 | 6 | 8 | 5 | 2 |
| 6 | 2 | 3 | 4 | 8 | 9 | 7 | 1 | 5 |
| 8 | 9 | 7 | 3 | 5 | 1 | 2 | 4 | 6 |
| 1 | 4 | 5 | 6 | 2 | 7 | 9 | 8 | 3 |

## Puzzle 13

| 8 | 4 | 7 | 2 | 1 | 9 | 3 | 6 | 5 |
|---|---|---|---|---|---|---|---|---|
| 3 | 1 | 5 | 7 | 8 | 6 | 2 | 9 | 4 |
| 9 | 6 | 2 | 4 | 5 | 3 | 1 | 7 | 8 |
| 5 | 2 | 1 | 9 | 7 | 4 | 6 | 8 | 3 |
| 6 | 8 | 4 | 3 | 2 | 5 | 7 | 1 | 9 |
| 7 | 3 | 9 | 1 | 6 | 8 | 5 | 4 | 2 |
| 2 | 9 | 3 | 6 | 4 | 7 | 8 | 5 | 1 |
| 1 | 5 | 6 | 8 | 9 | 2 | 4 | 3 | 7 |
| 4 | 7 | 8 | 5 | 3 | 1 | 9 | 2 | 6 |

## Puzzle 14

| 2 | 1 | 6 | 4 | 3 | 5 | 9 | 7 | 8 |
|---|---|---|---|---|---|---|---|---|
| 8 | 5 | 9 | 6 | 2 | 7 | 4 | 3 | 1 |
| 4 | 7 | 3 | 9 | 1 | 8 | 6 | 2 | 5 |
| 1 | 8 | 5 | 7 | 4 | 6 | 3 | 9 | 2 |
| 9 | 6 | 7 | 2 | 5 | 3 | 1 | 8 | 4 |
| 3 | 4 | 2 | 8 | 9 | 1 | 5 | 6 | 7 |
| 7 | 2 | 4 | 1 | 6 | 9 | 8 | 5 | 3 |
| 5 | 9 | 8 | 3 | 7 | 4 | 2 | 1 | 6 |
| 6 | 3 | 1 | 5 | 8 | 2 | 7 | 4 | 9 |

## Puzzle 15

| 4 | 2 | 6 | 3 | 1 | 5 | 8 | 7 | 9 |
|---|---|---|---|---|---|---|---|---|
| 8 | 5 | 1 | 4 | 7 | 9 | 2 | 6 | 3 |
| 7 | 9 | 3 | 2 | 8 | 6 | 1 | 4 | 5 |
| 3 | 4 | 9 | 7 | 2 | 8 | 6 | 5 | 1 |
| 2 | 7 | 5 | 1 | 6 | 3 | 9 | 8 | 4 |
| 6 | 1 | 8 | 5 | 9 | 4 | 3 | 2 | 7 |
| 9 | 3 | 7 | 6 | 4 | 2 | 5 | 1 | 8 |
| 5 | 6 | 4 | 8 | 3 | 1 | 7 | 9 | 2 |
| 1 | 8 | 2 | 9 | 5 | 7 | 4 | 3 | 6 |

## Puzzle 16

| 4 | 5 | 2 | 7 | 6 | 9 | 1 | 8 | 3 |
|---|---|---|---|---|---|---|---|---|
| 3 | 7 | 9 | 8 | 1 | 4 | 2 | 5 | 6 |
| 1 | 6 | 8 | 2 | 3 | 5 | 4 | 9 | 7 |
| 5 | 9 | 1 | 3 | 4 | 2 | 7 | 6 | 8 |
| 6 | 8 | 7 | 9 | 5 | 1 | 3 | 2 | 4 |
| 2 | 3 | 4 | 6 | 8 | 7 | 5 | 1 | 9 |
| 8 | 1 | 5 | 4 | 9 | 3 | 6 | 7 | 2 |
| 9 | 2 | 3 | 5 | 7 | 6 | 8 | 4 | 1 |
| 7 | 4 | 6 | 1 | 2 | 8 | 9 | 3 | 5 |

## Puzzle 17

| 4 | 5 | 1 | 2 | 9 | 7 | 6 | 8 | 3 |
|---|---|---|---|---|---|---|---|---|
| 2 | 8 | 9 | 6 | 3 | 1 | 5 | 4 | 7 |
| 6 | 7 | 3 | 5 | 8 | 4 | 9 | 1 | 2 |
| 5 | 9 | 4 | 7 | 1 | 3 | 8 | 2 | 6 |
| 1 | 6 | 2 | 4 | 5 | 8 | 7 | 3 | 9 |
| 8 | 3 | 7 | 9 | 6 | 2 | 1 | 5 | 4 |
| 3 | 2 | 8 | 1 | 7 | 9 | 4 | 6 | 5 |
| 7 | 1 | 5 | 3 | 4 | 6 | 2 | 9 | 8 |
| 9 | 4 | 6 | 8 | 2 | 5 | 3 | 7 | 1 |

## Puzzle 18

| 4 | 6 | 9 | 2 | 1 | 8 | 3 | 7 | 5 |
|---|---|---|---|---|---|---|---|---|
| 3 | 7 | 1 | 6 | 5 | 9 | 2 | 4 | 8 |
| 2 | 8 | 5 | 3 | 7 | 4 | 6 | 1 | 9 |
| 8 | 1 | 6 | 4 | 9 | 7 | 5 | 3 | 2 |
| 9 | 5 | 4 | 1 | 2 | 3 | 7 | 8 | 6 |
| 7 | 3 | 2 | 5 | 8 | 6 | 4 | 9 | 1 |
| 5 | 2 | 8 | 7 | 3 | 1 | 9 | 6 | 4 |
| 1 | 4 | 7 | 9 | 6 | 2 | 8 | 5 | 3 |
| 6 | 9 | 3 | 8 | 4 | 5 | 1 | 2 | 7 |

## Puzzle 19

| 6 | 8 | 9 | 1 | 3 | 4 | 7 | 5 | 2 |
|---|---|---|---|---|---|---|---|---|
| 2 | 7 | 4 | 8 | 6 | 5 | 3 | 1 | 9 |
| 5 | 1 | 3 | 9 | 7 | 2 | 4 | 6 | 8 |
| 1 | 9 | 6 | 3 | 2 | 7 | 5 | 8 | 4 |
| 4 | 2 | 7 | 5 | 9 | 8 | 6 | 3 | 1 |
| 3 | 5 | 8 | 4 | 1 | 6 | 2 | 9 | 7 |
| 7 | 6 | 1 | 2 | 5 | 9 | 8 | 4 | 3 |
| 9 | 4 | 5 | 7 | 8 | 3 | 1 | 2 | 6 |
| 8 | 3 | 2 | 6 | 4 | 1 | 9 | 7 | 5 |

## Puzzle 20

| 3 | 2 | 5 | 9 | 6 | 8 | 4 | 1 | 7 |
|---|---|---|---|---|---|---|---|---|
| 8 | 6 | 1 | 3 | 7 | 4 | 2 | 5 | 9 |
| 9 | 7 | 4 | 5 | 2 | 1 | 3 | 6 | 8 |
| 1 | 9 | 2 | 6 | 4 | 3 | 7 | 8 | 5 |
| 6 | 4 | 8 | 2 | 5 | 7 | 1 | 9 | 3 |
| 7 | 5 | 3 | 1 | 8 | 9 | 6 | 2 | 4 |
| 2 | 1 | 9 | 4 | 3 | 5 | 8 | 7 | 6 |
| 4 | 8 | 6 | 7 | 9 | 2 | 5 | 3 | 1 |
| 5 | 3 | 7 | 8 | 1 | 6 | 9 | 4 | 2 |

## Puzzle 21

| 7 | 1 | 9 | 2 | 4 | 8 | 3 | 5 | 6 |
|---|---|---|---|---|---|---|---|---|
| 8 | 5 | 3 | 1 | 6 | 7 | 2 | 4 | 9 |
| 6 | 4 | 2 | 9 | 3 | 5 | 7 | 1 | 8 |
| 5 | 6 | 7 | 8 | 2 | 4 | 1 | 9 | 3 |
| 9 | 2 | 1 | 5 | 7 | 3 | 8 | 6 | 4 |
| 4 | 3 | 8 | 6 | 1 | 9 | 5 | 7 | 2 |
| 1 | 9 | 5 | 4 | 8 | 2 | 6 | 3 | 7 |
| 2 | 7 | 4 | 3 | 5 | 6 | 9 | 8 | 1 |
| 3 | 8 | 6 | 7 | 9 | 1 | 4 | 2 | 5 |

## Puzzle 22

| 6 | 2 | 4 | 7 | 5 | 9 | 3 | 1 | 8 |
|---|---|---|---|---|---|---|---|---|
| 7 | 1 | 8 | 4 | 6 | 3 | 9 | 5 | 2 |
| 3 | 5 | 9 | 8 | 2 | 1 | 4 | 6 | 7 |
| 5 | 7 | 1 | 2 | 3 | 4 | 8 | 9 | 6 |
| 4 | 8 | 2 | 6 | 9 | 5 | 7 | 3 | 1 |
| 9 | 6 | 3 | 1 | 8 | 7 | 2 | 4 | 5 |
| 2 | 3 | 5 | 9 | 7 | 6 | 1 | 8 | 4 |
| 1 | 9 | 7 | 5 | 4 | 8 | 6 | 2 | 3 |
| 8 | 4 | 6 | 3 | 1 | 2 | 5 | 7 | 9 |

## Puzzle 23

| 7 | 4 | 8 | 5 | 6 | 9 | 1 | 3 | 2 |
|---|---|---|---|---|---|---|---|---|
| 3 | 1 | 9 | 2 | 4 | 7 | 5 | 8 | 6 |
| 5 | 2 | 6 | 1 | 8 | 3 | 9 | 4 | 7 |
| 6 | 8 | 1 | 7 | 9 | 4 | 2 | 5 | 3 |
| 9 | 3 | 4 | 8 | 2 | 5 | 7 | 6 | 1 |
| 2 | 7 | 5 | 3 | 1 | 6 | 8 | 9 | 4 |
| 4 | 5 | 3 | 9 | 7 | 1 | 6 | 2 | 8 |
| 8 | 6 | 7 | 4 | 5 | 2 | 3 | 1 | 9 |
| 1 | 9 | 2 | 6 | 3 | 8 | 4 | 7 | 5 |

## Puzzle 24

| 3 | 4 | 9 | 5 | 7 | 6 | 1 | 2 | 8 |
|---|---|---|---|---|---|---|---|---|
| 7 | 6 | 1 | 4 | 2 | 8 | 9 | 5 | 3 |
| 2 | 8 | 5 | 1 | 3 | 9 | 4 | 7 | 6 |
| 1 | 3 | 7 | 6 | 4 | 5 | 2 | 8 | 9 |
| 8 | 2 | 6 | 3 | 9 | 1 | 5 | 4 | 7 |
| 9 | 5 | 4 | 2 | 8 | 7 | 6 | 3 | 1 |
| 6 | 7 | 8 | 9 | 5 | 4 | 3 | 1 | 2 |
| 5 | 9 | 2 | 7 | 1 | 3 | 8 | 6 | 4 |
| 4 | 1 | 3 | 8 | 6 | 2 | 7 | 9 | 5 |

## Puzzle 25

| 8 | 5 | 1 | 4 | 9 | 3 | 6 | 7 | 2 |
|---|---|---|---|---|---|---|---|---|
| 3 | 7 | 4 | 2 | 1 | 6 | 8 | 5 | 9 |
| 6 | 9 | 2 | 8 | 5 | 7 | 3 | 4 | 1 |
| 5 | 1 | 9 | 3 | 8 | 2 | 7 | 6 | 4 |
| 4 | 6 | 8 | 1 | 7 | 9 | 5 | 2 | 3 |
| 2 | 3 | 7 | 6 | 4 | 5 | 9 | 1 | 8 |
| 7 | 2 | 3 | 9 | 6 | 4 | 1 | 8 | 5 |
| 1 | 4 | 5 | 7 | 3 | 8 | 2 | 9 | 6 |
| 9 | 8 | 6 | 5 | 2 | 1 | 4 | 3 | 7 |

## Puzzle 26

| 8 | 9 | 1 | 3 | 2 | 4 | 7 | 6 | 5 |
|---|---|---|---|---|---|---|---|---|
| 7 | 5 | 4 | 6 | 9 | 8 | 1 | 2 | 3 |
| 3 | 2 | 6 | 7 | 1 | 5 | 8 | 4 | 9 |
| 1 | 4 | 7 | 9 | 5 | 3 | 2 | 8 | 6 |
| 6 | 8 | 9 | 2 | 4 | 7 | 3 | 5 | 1 |
| 2 | 3 | 5 | 1 | 8 | 6 | 4 | 9 | 7 |
| 4 | 7 | 8 | 5 | 3 | 9 | 6 | 1 | 2 |
| 9 | 1 | 3 | 8 | 6 | 2 | 5 | 7 | 4 |
| 5 | 6 | 2 | 4 | 7 | 1 | 9 | 3 | 8 |

## Puzzle 27

| 1 | 5 | 4 | 7 | 2 | 8 | 9 | 3 | 6 |
|---|---|---|---|---|---|---|---|---|
| 3 | 2 | 8 | 6 | 9 | 5 | 7 | 1 | 4 |
| 7 | 9 | 6 | 4 | 1 | 3 | 2 | 8 | 5 |
| 8 | 3 | 7 | 1 | 4 | 6 | 5 | 2 | 9 |
| 9 | 1 | 5 | 2 | 3 | 7 | 4 | 6 | 8 |
| 6 | 4 | 2 | 8 | 5 | 9 | 3 | 7 | 1 |
| 4 | 6 | 9 | 3 | 7 | 1 | 8 | 5 | 2 |
| 5 | 7 | 1 | 9 | 8 | 2 | 6 | 4 | 3 |
| 2 | 8 | 3 | 5 | 6 | 4 | 1 | 9 | 7 |

## Puzzle 28

| 6 | 9 | 5 | 3 | 8 | 7 | 1 | 4 | 2 |
|---|---|---|---|---|---|---|---|---|
| 3 | 2 | 1 | 5 | 6 | 4 | 9 | 8 | 7 |
| 4 | 7 | 8 | 9 | 1 | 2 | 6 | 5 | 3 |
| 8 | 1 | 3 | 2 | 7 | 9 | 5 | 6 | 4 |
| 9 | 6 | 2 | 4 | 5 | 8 | 7 | 3 | 1 |
| 5 | 4 | 7 | 6 | 3 | 1 | 8 | 2 | 9 |
| 2 | 5 | 9 | 7 | 4 | 6 | 3 | 1 | 8 |
| 7 | 8 | 6 | 1 | 2 | 3 | 4 | 9 | 5 |
| 1 | 3 | 4 | 8 | 9 | 5 | 2 | 7 | 6 |

## Puzzle 29

| 2 | 3 | 4 | 7 | 6 | 8 | 1 | 9 | 5 |
|---|---|---|---|---|---|---|---|---|
| 5 | 6 | 1 | 9 | 4 | 2 | 3 | 8 | 7 |
| 8 | 7 | 9 | 3 | 5 | 1 | 4 | 2 | 6 |
| 7 | 8 | 6 | 2 | 9 | 3 | 5 | 1 | 4 |
| 9 | 5 | 3 | 6 | 1 | 4 | 8 | 7 | 2 |
| 4 | 1 | 2 | 5 | 8 | 7 | 6 | 3 | 9 |
| 1 | 9 | 8 | 4 | 7 | 6 | 2 | 5 | 3 |
| 3 | 4 | 5 | 8 | 2 | 9 | 7 | 6 | 1 |
| 6 | 2 | 7 | 1 | 3 | 5 | 9 | 4 | 8 |

## Puzzle 30

| 1 | 4 | 8 | 2 | 7 | 3 | 6 | 5 | 9 |
|---|---|---|---|---|---|---|---|---|
| 9 | 3 | 5 | 4 | 8 | 6 | 2 | 1 | 7 |
| 7 | 6 | 2 | 1 | 9 | 5 | 3 | 4 | 8 |
| 6 | 1 | 9 | 8 | 3 | 2 | 4 | 7 | 5 |
| 5 | 7 | 3 | 6 | 4 | 9 | 1 | 8 | 2 |
| 8 | 2 | 4 | 5 | 1 | 7 | 9 | 6 | 3 |
| 3 | 5 | 1 | 7 | 2 | 4 | 8 | 9 | 6 |
| 4 | 9 | 6 | 3 | 5 | 8 | 7 | 2 | 1 |
| 2 | 8 | 7 | 9 | 6 | 1 | 5 | 3 | 4 |

## Puzzle 31

| 7 | 6 | 1 | 4 | 2 | 3 | 5 | 9 | 8 |
|---|---|---|---|---|---|---|---|---|
| 9 | 8 | 2 | 7 | 6 | 5 | 1 | 4 | 3 |
| 4 | 3 | 5 | 8 | 1 | 9 | 7 | 2 | 6 |
| 1 | 4 | 7 | 3 | 8 | 2 | 9 | 6 | 5 |
| 8 | 2 | 9 | 6 | 5 | 1 | 4 | 3 | 7 |
| 3 | 5 | 6 | 9 | 4 | 7 | 2 | 8 | 1 |
| 5 | 7 | 4 | 2 | 3 | 8 | 6 | 1 | 9 |
| 2 | 9 | 8 | 1 | 7 | 6 | 3 | 5 | 4 |
| 6 | 1 | 3 | 5 | 9 | 4 | 8 | 7 | 2 |

## Puzzle 32

| 3 | 4 | 1 | 5 | 9 | 6 | 8 | 2 | 7 |
|---|---|---|---|---|---|---|---|---|
| 7 | 6 | 2 | 8 | 3 | 1 | 4 | 5 | 9 |
| 8 | 5 | 9 | 4 | 7 | 2 | 6 | 1 | 3 |
| 5 | 8 | 6 | 7 | 1 | 9 | 3 | 4 | 2 |
| 1 | 2 | 4 | 6 | 5 | 3 | 9 | 7 | 8 |
| 9 | 3 | 7 | 2 | 4 | 8 | 5 | 6 | 1 |
| 2 | 7 | 3 | 9 | 6 | 4 | 1 | 8 | 5 |
| 4 | 9 | 8 | 1 | 2 | 5 | 7 | 3 | 6 |
| 6 | 1 | 5 | 3 | 8 | 7 | 2 | 9 | 4 |

## Puzzle 33

| 3 | 5 | 7 | 2 | 1 | 4 | 8 | 6 | 9 |
|---|---|---|---|---|---|---|---|---|
| 1 | 6 | 4 | 5 | 8 | 9 | 2 | 7 | 3 |
| 2 | 8 | 9 | 6 | 7 | 3 | 4 | 5 | 1 |
| 6 | 4 | 8 | 7 | 9 | 2 | 1 | 3 | 5 |
| 9 | 1 | 2 | 3 | 6 | 5 | 7 | 8 | 4 |
| 5 | 7 | 3 | 8 | 4 | 1 | 6 | 9 | 2 |
| 4 | 3 | 6 | 1 | 5 | 7 | 9 | 2 | 8 |
| 8 | 2 | 1 | 9 | 3 | 6 | 5 | 4 | 7 |
| 7 | 9 | 5 | 4 | 2 | 8 | 3 | 1 | 6 |

## Puzzle 34

| 6 | 5 | 3 | 1 | 9 | 8 | 2 | 4 | 7 |
|---|---|---|---|---|---|---|---|---|
| 8 | 1 | 9 | 2 | 7 | 4 | 6 | 3 | 5 |
| 4 | 2 | 7 | 5 | 3 | 6 | 9 | 8 | 1 |
| 7 | 3 | 1 | 9 | 8 | 5 | 4 | 6 | 2 |
| 5 | 9 | 6 | 4 | 1 | 2 | 3 | 7 | 8 |
| 2 | 4 | 8 | 3 | 6 | 7 | 5 | 1 | 9 |
| 3 | 6 | 2 | 8 | 5 | 1 | 7 | 9 | 4 |
| 1 | 7 | 5 | 6 | 4 | 9 | 8 | 2 | 3 |
| 9 | 8 | 4 | 7 | 2 | 3 | 1 | 5 | 6 |

## Puzzle 35

| 5 | 1 | 9 | 4 | 6 | 8 | 2 | 7 | 3 |
|---|---|---|---|---|---|---|---|---|
| 6 | 2 | 4 | 1 | 7 | 3 | 9 | 8 | 5 |
| 8 | 7 | 3 | 5 | 9 | 2 | 6 | 1 | 4 |
| 3 | 6 | 5 | 8 | 1 | 9 | 4 | 2 | 7 |
| 9 | 8 | 2 | 7 | 4 | 6 | 5 | 3 | 1 |
| 1 | 4 | 7 | 2 | 3 | 5 | 8 | 9 | 6 |
| 4 | 3 | 8 | 9 | 5 | 1 | 7 | 6 | 2 |
| 7 | 9 | 1 | 6 | 2 | 4 | 3 | 5 | 8 |
| 2 | 5 | 6 | 3 | 8 | 7 | 1 | 4 | 9 |

## Puzzle 36

| 1 | 2 | 7 | 3 | 8 | 4 | 6 | 9 | 5 |
|---|---|---|---|---|---|---|---|---|
| 5 | 4 | 8 | 9 | 7 | 6 | 2 | 3 | 1 |
| 6 | 3 | 9 | 5 | 1 | 2 | 8 | 7 | 4 |
| 7 | 8 | 4 | 2 | 9 | 1 | 3 | 5 | 6 |
| 9 | 6 | 2 | 4 | 5 | 3 | 7 | 1 | 8 |
| 3 | 1 | 5 | 7 | 6 | 8 | 4 | 2 | 9 |
| 4 | 7 | 6 | 1 | 3 | 5 | 9 | 8 | 2 |
| 2 | 9 | 1 | 8 | 4 | 7 | 5 | 6 | 3 |
| 8 | 5 | 3 | 6 | 2 | 9 | 1 | 4 | 7 |

## Puzzle 37

| 3 | 4 | 6 | 7 | 5 | 8 | 1 | 9 | 2 |
|---|---|---|---|---|---|---|---|---|
| 8 | 2 | 1 | 9 | 3 | 6 | 7 | 5 | 4 |
| 5 | 7 | 9 | 4 | 1 | 2 | 3 | 8 | 6 |
| 9 | 8 | 7 | 1 | 2 | 5 | 6 | 4 | 3 |
| 4 | 5 | 3 | 6 | 7 | 9 | 2 | 1 | 8 |
| 6 | 1 | 2 | 8 | 4 | 3 | 5 | 7 | 9 |
| 7 | 9 | 5 | 3 | 6 | 4 | 8 | 2 | 1 |
| 1 | 3 | 4 | 2 | 8 | 7 | 9 | 6 | 5 |
| 2 | 6 | 8 | 5 | 9 | 1 | 4 | 3 | 7 |

## Puzzle 38

| 6 | 3 | 7 | 4 | 8 | 2 | 1 | 5 | 9 |
|---|---|---|---|---|---|---|---|---|
| 1 | 8 | 5 | 9 | 6 | 3 | 7 | 2 | 4 |
| 9 | 2 | 4 | 1 | 5 | 7 | 3 | 6 | 8 |
| 8 | 5 | 1 | 7 | 3 | 6 | 4 | 9 | 2 |
| 2 | 4 | 3 | 5 | 1 | 9 | 6 | 8 | 7 |
| 7 | 9 | 6 | 2 | 4 | 8 | 5 | 3 | 1 |
| 5 | 7 | 8 | 3 | 9 | 1 | 2 | 4 | 6 |
| 4 | 6 | 2 | 8 | 7 | 5 | 9 | 1 | 3 |
| 3 | 1 | 9 | 6 | 2 | 4 | 8 | 7 | 5 |

## Puzzle 39

| 4 | 9 | 7 | 3 | 1 | 8 | 2 | 5 | 6 |
|---|---|---|---|---|---|---|---|---|
| 5 | 3 | 1 | 7 | 6 | 2 | 9 | 8 | 4 |
| 6 | 2 | 8 | 9 | 4 | 5 | 1 | 7 | 3 |
| 1 | 6 | 2 | 8 | 7 | 3 | 4 | 9 | 5 |
| 8 | 5 | 3 | 1 | 9 | 4 | 6 | 2 | 7 |
| 7 | 4 | 9 | 5 | 2 | 6 | 3 | 1 | 8 |
| 2 | 7 | 5 | 6 | 3 | 9 | 8 | 4 | 1 |
| 3 | 8 | 4 | 2 | 5 | 1 | 7 | 6 | 9 |
| 9 | 1 | 6 | 4 | 8 | 7 | 5 | 3 | 2 |

## Puzzle 40

| 5 | 4 | 6 | 7 | 3 | 1 | 8 | 2 | 9 |
|---|---|---|---|---|---|---|---|---|
| 2 | 1 | 3 | 9 | 8 | 6 | 4 | 7 | 5 |
| 9 | 8 | 7 | 4 | 5 | 2 | 6 | 1 | 3 |
| 6 | 9 | 1 | 8 | 2 | 3 | 5 | 4 | 7 |
| 3 | 7 | 8 | 5 | 4 | 9 | 2 | 6 | 1 |
| 4 | 5 | 2 | 6 | 1 | 7 | 3 | 9 | 8 |
| 8 | 2 | 4 | 1 | 9 | 5 | 7 | 3 | 6 |
| 1 | 6 | 5 | 3 | 7 | 4 | 9 | 8 | 2 |
| 7 | 3 | 9 | 2 | 6 | 8 | 1 | 5 | 4 |

## Puzzle 41

| 9 | 2 | 1 | 4 | 6 | 5 | 3 | 8 | 7 |
|---|---|---|---|---|---|---|---|---|
| 8 | 4 | 7 | 3 | 9 | 2 | 1 | 5 | 6 |
| 5 | 3 | 6 | 7 | 1 | 8 | 9 | 2 | 4 |
| 7 | 1 | 4 | 9 | 2 | 3 | 5 | 6 | 8 |
| 2 | 5 | 9 | 8 | 4 | 6 | 7 | 3 | 1 |
| 3 | 6 | 8 | 5 | 7 | 1 | 2 | 4 | 9 |
| 4 | 9 | 3 | 2 | 8 | 7 | 6 | 1 | 5 |
| 6 | 8 | 2 | 1 | 5 | 9 | 4 | 7 | 3 |
| 1 | 7 | 5 | 6 | 3 | 4 | 8 | 9 | 2 |

## Puzzle 42

| 3 | 2 | 4 | 8 | 6 | 5 | 1 | 9 | 7 |
|---|---|---|---|---|---|---|---|---|
| 5 | 6 | 9 | 7 | 2 | 1 | 3 | 8 | 4 |
| 8 | 7 | 1 | 3 | 9 | 4 | 6 | 2 | 5 |
| 6 | 1 | 2 | 5 | 4 | 7 | 8 | 3 | 9 |
| 4 | 9 | 5 | 6 | 3 | 8 | 2 | 7 | 1 |
| 7 | 8 | 3 | 2 | 1 | 9 | 4 | 5 | 6 |
| 2 | 5 | 8 | 4 | 7 | 6 | 9 | 1 | 3 |
| 9 | 3 | 6 | 1 | 5 | 2 | 7 | 4 | 8 |
| 1 | 4 | 7 | 9 | 8 | 3 | 5 | 6 | 2 |

## Puzzle 43

| 3 | 4 | 9 | 8 | 7 | 1 | 5 | 2 | 6 |
|---|---|---|---|---|---|---|---|---|
| 1 | 5 | 2 | 6 | 3 | 4 | 9 | 8 | 7 |
| 6 | 8 | 7 | 9 | 2 | 5 | 1 | 3 | 4 |
| 7 | 6 | 4 | 5 | 1 | 3 | 2 | 9 | 8 |
| 9 | 2 | 5 | 7 | 4 | 8 | 6 | 1 | 3 |
| 8 | 3 | 1 | 2 | 6 | 9 | 4 | 7 | 5 |
| 2 | 7 | 8 | 4 | 9 | 6 | 3 | 5 | 1 |
| 5 | 1 | 6 | 3 | 8 | 2 | 7 | 4 | 9 |
| 4 | 9 | 3 | 1 | 5 | 7 | 8 | 6 | 2 |

## Puzzle 44

| 1 | 7 | 6 | 3 | 2 | 9 | 5 | 4 | 8 |
|---|---|---|---|---|---|---|---|---|
| 3 | 9 | 8 | 1 | 5 | 4 | 2 | 6 | 7 |
| 5 | 2 | 4 | 6 | 7 | 8 | 3 | 1 | 9 |
| 4 | 5 | 1 | 8 | 3 | 6 | 9 | 7 | 2 |
| 9 | 3 | 2 | 5 | 1 | 7 | 4 | 8 | 6 |
| 6 | 8 | 7 | 4 | 9 | 2 | 1 | 3 | 5 |
| 8 | 1 | 3 | 9 | 6 | 5 | 7 | 2 | 4 |
| 7 | 4 | 9 | 2 | 8 | 1 | 6 | 5 | 3 |
| 2 | 6 | 5 | 7 | 4 | 3 | 8 | 9 | 1 |

## Puzzle 45

| 1 | 9 | 5 | 4 | 2 | 7 | 8 | 3 | 6 |
|---|---|---|---|---|---|---|---|---|
| 8 | 7 | 6 | 5 | 3 | 9 | 1 | 2 | 4 |
| 3 | 2 | 4 | 8 | 6 | 1 | 9 | 5 | 7 |
| 6 | 1 | 7 | 2 | 9 | 5 | 4 | 8 | 3 |
| 4 | 8 | 9 | 6 | 7 | 3 | 2 | 1 | 5 |
| 5 | 3 | 2 | 1 | 8 | 4 | 6 | 7 | 9 |
| 2 | 5 | 3 | 9 | 1 | 6 | 7 | 4 | 8 |
| 7 | 6 | 8 | 3 | 4 | 2 | 5 | 9 | 1 |
| 9 | 4 | 1 | 7 | 5 | 8 | 3 | 6 | 2 |

## Puzzle 46

| 2 | 8 | 6 | 1 | 4 | 9 | 7 | 3 | 5 |
|---|---|---|---|---|---|---|---|---|
| 7 | 9 | 5 | 8 | 2 | 3 | 6 | 1 | 4 |
| 3 | 1 | 4 | 7 | 5 | 6 | 2 | 8 | 9 |
| 8 | 2 | 1 | 4 | 9 | 7 | 5 | 6 | 3 |
| 5 | 3 | 9 | 6 | 1 | 2 | 4 | 7 | 8 |
| 6 | 4 | 7 | 5 | 3 | 8 | 1 | 9 | 2 |
| 9 | 6 | 2 | 3 | 7 | 4 | 8 | 5 | 1 |
| 1 | 7 | 3 | 2 | 8 | 5 | 9 | 4 | 6 |
| 4 | 5 | 8 | 9 | 6 | 1 | 3 | 2 | 7 |

## Puzzle 47

| 7 | 5 | 8 | 2 | 9 | 4 | 3 | 6 | 1 |
|---|---|---|---|---|---|---|---|---|
| 6 | 1 | 2 | 8 | 7 | 3 | 4 | 5 | 9 |
| 3 | 4 | 9 | 5 | 1 | 6 | 7 | 2 | 8 |
| 8 | 6 | 4 | 3 | 2 | 5 | 9 | 1 | 7 |
| 9 | 3 | 5 | 7 | 4 | 1 | 6 | 8 | 2 |
| 2 | 7 | 1 | 6 | 8 | 9 | 5 | 4 | 3 |
| 5 | 2 | 3 | 9 | 6 | 8 | 1 | 7 | 4 |
| 1 | 8 | 6 | 4 | 3 | 7 | 2 | 9 | 5 |
| 4 | 9 | 7 | 1 | 5 | 2 | 8 | 3 | 6 |

## Puzzle 48

| 9 | 8 | 3 | 6 | 4 | 7 | 5 | 2 | 1 |
|---|---|---|---|---|---|---|---|---|
| 4 | 2 | 6 | 8 | 1 | 5 | 3 | 7 | 9 |
| 1 | 5 | 7 | 3 | 2 | 9 | 8 | 4 | 6 |
| 5 | 6 | 9 | 4 | 8 | 1 | 7 | 3 | 2 |
| 3 | 1 | 4 | 7 | 5 | 2 | 6 | 9 | 8 |
| 8 | 7 | 2 | 9 | 6 | 3 | 4 | 1 | 5 |
| 2 | 3 | 1 | 5 | 7 | 8 | 9 | 6 | 4 |
| 7 | 4 | 5 | 2 | 9 | 6 | 1 | 8 | 3 |
| 6 | 9 | 8 | 1 | 3 | 4 | 2 | 5 | 7 |

## Puzzle 49

| 7 | 8 | 6 | 5 | 1 | 9 | 4 | 2 | 3 |
|---|---|---|---|---|---|---|---|---|
| 9 | 4 | 5 | 8 | 2 | 3 | 1 | 6 | 7 |
| 2 | 1 | 3 | 6 | 4 | 7 | 9 | 8 | 5 |
| 8 | 3 | 2 | 4 | 5 | 1 | 7 | 9 | 6 |
| 1 | 9 | 7 | 2 | 3 | 6 | 5 | 4 | 8 |
| 5 | 6 | 4 | 9 | 7 | 8 | 3 | 1 | 2 |
| 3 | 7 | 8 | 1 | 6 | 4 | 2 | 5 | 9 |
| 4 | 2 | 9 | 3 | 8 | 5 | 6 | 7 | 1 |
| 6 | 5 | 1 | 7 | 9 | 2 | 8 | 3 | 4 |

## Puzzle 50

| 4 | 1 | 2 | 3 | 9 | 8 | 5 | 6 | 7 |
|---|---|---|---|---|---|---|---|---|
| 9 | 3 | 8 | 5 | 6 | 7 | 2 | 1 | 4 |
| 7 | 5 | 6 | 1 | 4 | 2 | 3 | 8 | 9 |
| 3 | 6 | 9 | 4 | 8 | 5 | 1 | 7 | 2 |
| 8 | 2 | 4 | 7 | 1 | 3 | 6 | 9 | 5 |
| 1 | 7 | 5 | 6 | 2 | 9 | 4 | 3 | 8 |
| 5 | 8 | 7 | 2 | 3 | 1 | 9 | 4 | 6 |
| 2 | 4 | 1 | 9 | 7 | 6 | 8 | 5 | 3 |
| 6 | 9 | 3 | 8 | 5 | 4 | 7 | 2 | 1 |

## Puzzle 51

| 4 | 2 | 7 | 8 | 5 | 3 | 9 | 6 | 1 |
|---|---|---|---|---|---|---|---|---|
| 6 | 3 | 5 | 9 | 1 | 4 | 2 | 8 | 7 |
| 1 | 8 | 9 | 2 | 6 | 7 | 5 | 3 | 4 |
| 8 | 6 | 3 | 5 | 4 | 9 | 1 | 7 | 2 |
| 7 | 5 | 1 | 6 | 3 | 2 | 8 | 4 | 9 |
| 2 | 9 | 4 | 1 | 7 | 8 | 6 | 5 | 3 |
| 9 | 7 | 6 | 3 | 2 | 5 | 4 | 1 | 8 |
| 3 | 1 | 8 | 4 | 9 | 6 | 7 | 2 | 5 |
| 5 | 4 | 2 | 7 | 8 | 1 | 3 | 9 | 6 |

## Puzzle 52

| 3 | 4 | 7 | 2 | 1 | 6 | 5 | 9 | 8 |
|---|---|---|---|---|---|---|---|---|
| 1 | 5 | 6 | 3 | 9 | 8 | 2 | 7 | 4 |
| 2 | 9 | 8 | 7 | 4 | 5 | 6 | 1 | 3 |
| 5 | 3 | 9 | 6 | 8 | 7 | 4 | 2 | 1 |
| 4 | 7 | 1 | 5 | 2 | 9 | 8 | 3 | 6 |
| 8 | 6 | 2 | 1 | 3 | 4 | 9 | 5 | 7 |
| 6 | 2 | 4 | 9 | 7 | 3 | 1 | 8 | 5 |
| 9 | 8 | 3 | 4 | 5 | 1 | 7 | 6 | 2 |
| 7 | 1 | 5 | 8 | 6 | 2 | 3 | 4 | 9 |

## Puzzle 53

| 8 | 4 | 3 | 2 | 5 | 9 | 7 | 1 | 6 |
|---|---|---|---|---|---|---|---|---|
| 5 | 6 | 1 | 7 | 4 | 8 | 9 | 3 | 2 |
| 7 | 9 | 2 | 6 | 1 | 3 | 4 | 8 | 5 |
| 6 | 3 | 7 | 9 | 2 | 4 | 8 | 5 | 1 |
| 9 | 5 | 4 | 1 | 8 | 6 | 3 | 2 | 7 |
| 1 | 2 | 8 | 5 | 3 | 7 | 6 | 9 | 4 |
| 2 | 7 | 9 | 8 | 6 | 1 | 5 | 4 | 3 |
| 4 | 8 | 5 | 3 | 7 | 2 | 1 | 6 | 9 |
| 3 | 1 | 6 | 4 | 9 | 5 | 2 | 7 | 8 |

## Puzzle 54

| 7 | 9 | 6 | 1 | 3 | 4 | 2 | 5 | 8 |
|---|---|---|---|---|---|---|---|---|
| 2 | 3 | 1 | 8 | 5 | 7 | 6 | 9 | 4 |
| 8 | 5 | 4 | 6 | 2 | 9 | 1 | 7 | 3 |
| 3 | 2 | 8 | 7 | 9 | 6 | 4 | 1 | 5 |
| 9 | 4 | 7 | 3 | 1 | 5 | 8 | 6 | 2 |
| 1 | 6 | 5 | 4 | 8 | 2 | 7 | 3 | 9 |
| 5 | 8 | 2 | 9 | 7 | 1 | 3 | 4 | 6 |
| 4 | 7 | 3 | 5 | 6 | 8 | 9 | 2 | 1 |
| 6 | 1 | 9 | 2 | 4 | 3 | 5 | 8 | 7 |

## Puzzle 55

| 9 | 6 | 4 | 1 | 7 | 2 | 8 | 5 | 3 |
| 3 | 2 | 8 | 5 | 6 | 4 | 9 | 1 | 7 |
| 1 | 7 | 5 | 9 | 8 | 3 | 6 | 4 | 2 |
| 5 | 3 | 7 | 2 | 4 | 9 | 1 | 8 | 6 |
| 4 | 9 | 1 | 6 | 3 | 8 | 2 | 7 | 5 |
| 6 | 8 | 2 | 7 | 5 | 1 | 3 | 9 | 4 |
| 8 | 5 | 6 | 3 | 9 | 7 | 4 | 2 | 1 |
| 2 | 4 | 3 | 8 | 1 | 5 | 7 | 6 | 9 |
| 7 | 1 | 9 | 4 | 2 | 6 | 5 | 3 | 8 |

## Puzzle 56

| 6 | 8 | 3 | 7 | 5 | 2 | 9 | 4 | 1 |
| 5 | 9 | 1 | 3 | 8 | 4 | 2 | 7 | 6 |
| 4 | 2 | 7 | 1 | 9 | 6 | 5 | 8 | 3 |
| 1 | 7 | 2 | 8 | 6 | 9 | 3 | 5 | 4 |
| 3 | 4 | 6 | 5 | 1 | 7 | 8 | 2 | 9 |
| 8 | 5 | 9 | 4 | 2 | 3 | 1 | 6 | 7 |
| 7 | 1 | 4 | 2 | 3 | 5 | 6 | 9 | 8 |
| 2 | 6 | 8 | 9 | 4 | 1 | 7 | 3 | 5 |
| 9 | 3 | 5 | 6 | 7 | 8 | 4 | 1 | 2 |

## Puzzle 57

| 3 | 8 | 9 | 5 | 6 | 1 | 7 | 4 | 2 |
| 6 | 2 | 7 | 3 | 8 | 4 | 5 | 1 | 9 |
| 1 | 5 | 4 | 9 | 2 | 7 | 6 | 8 | 3 |
| 9 | 4 | 6 | 7 | 1 | 5 | 2 | 3 | 8 |
| 8 | 7 | 2 | 6 | 4 | 3 | 9 | 5 | 1 |
| 5 | 1 | 3 | 2 | 9 | 8 | 4 | 7 | 6 |
| 2 | 3 | 8 | 4 | 5 | 9 | 1 | 6 | 7 |
| 4 | 6 | 1 | 8 | 7 | 2 | 3 | 9 | 5 |
| 7 | 9 | 5 | 1 | 3 | 6 | 8 | 2 | 4 |

## Puzzle 58

| 3 | 7 | 4 | 6 | 1 | 9 | 2 | 8 | 5 |
| 6 | 2 | 1 | 5 | 4 | 8 | 7 | 3 | 9 |
| 5 | 9 | 8 | 3 | 7 | 2 | 1 | 6 | 4 |
| 7 | 1 | 2 | 4 | 3 | 6 | 5 | 9 | 8 |
| 4 | 5 | 9 | 8 | 2 | 7 | 3 | 1 | 6 |
| 8 | 3 | 6 | 1 | 9 | 5 | 4 | 7 | 2 |
| 1 | 8 | 7 | 2 | 6 | 4 | 9 | 5 | 3 |
| 2 | 6 | 3 | 9 | 5 | 1 | 8 | 4 | 7 |
| 9 | 4 | 5 | 7 | 8 | 3 | 6 | 2 | 1 |

## Puzzle 59

| 5 | 3 | 7 | 4 | 8 | 2 | 9 | 6 | 1 |
| 4 | 6 | 2 | 9 | 3 | 1 | 8 | 7 | 5 |
| 1 | 9 | 8 | 6 | 7 | 5 | 4 | 2 | 3 |
| 2 | 1 | 5 | 7 | 9 | 8 | 6 | 3 | 4 |
| 3 | 7 | 6 | 1 | 5 | 4 | 2 | 9 | 8 |
| 8 | 4 | 9 | 2 | 6 | 3 | 5 | 1 | 7 |
| 7 | 5 | 4 | 3 | 2 | 9 | 1 | 8 | 6 |
| 6 | 2 | 1 | 8 | 4 | 7 | 3 | 5 | 9 |
| 9 | 8 | 3 | 5 | 1 | 6 | 7 | 4 | 2 |

## Puzzle 60

| 3 | 9 | 4 | 5 | 6 | 2 | 8 | 7 | 1 |
| 2 | 1 | 6 | 8 | 9 | 7 | 5 | 4 | 3 |
| 8 | 7 | 5 | 4 | 3 | 1 | 6 | 9 | 2 |
| 1 | 2 | 9 | 6 | 4 | 5 | 7 | 3 | 8 |
| 6 | 3 | 8 | 7 | 1 | 9 | 2 | 5 | 4 |
| 5 | 4 | 7 | 3 | 2 | 8 | 1 | 6 | 9 |
| 4 | 8 | 2 | 9 | 7 | 6 | 3 | 1 | 5 |
| 7 | 5 | 3 | 1 | 8 | 4 | 9 | 2 | 6 |
| 9 | 6 | 1 | 2 | 5 | 3 | 4 | 8 | 7 |

## Puzzle 61

| 7 | 8 | 4 | 6 | 5 | 2 | 3 | 9 | 1 |
|---|---|---|---|---|---|---|---|---|
| 9 | 5 | 3 | 4 | 8 | 1 | 7 | 6 | 2 |
| 6 | 2 | 1 | 9 | 7 | 3 | 5 | 8 | 4 |
| 3 | 6 | 7 | 8 | 2 | 4 | 1 | 5 | 9 |
| 5 | 1 | 8 | 3 | 9 | 6 | 2 | 4 | 7 |
| 2 | 4 | 9 | 7 | 1 | 5 | 6 | 3 | 8 |
| 1 | 3 | 2 | 5 | 4 | 8 | 9 | 7 | 6 |
| 4 | 9 | 6 | 1 | 3 | 7 | 8 | 2 | 5 |
| 8 | 7 | 5 | 2 | 6 | 9 | 4 | 1 | 3 |

## Puzzle 62

| 9 | 3 | 5 | 7 | 1 | 8 | 4 | 6 | 2 |
|---|---|---|---|---|---|---|---|---|
| 1 | 7 | 8 | 2 | 4 | 6 | 5 | 3 | 9 |
| 4 | 2 | 6 | 3 | 9 | 5 | 7 | 8 | 1 |
| 2 | 6 | 9 | 1 | 3 | 4 | 8 | 7 | 5 |
| 7 | 5 | 3 | 6 | 8 | 2 | 1 | 9 | 4 |
| 8 | 1 | 4 | 9 | 5 | 7 | 3 | 2 | 6 |
| 3 | 9 | 2 | 5 | 7 | 1 | 6 | 4 | 8 |
| 5 | 8 | 7 | 4 | 6 | 9 | 2 | 1 | 3 |
| 6 | 4 | 1 | 8 | 2 | 3 | 9 | 5 | 7 |

## Puzzle 63

| 2 | 7 | 6 | 4 | 8 | 9 | 5 | 1 | 3 |
|---|---|---|---|---|---|---|---|---|
| 5 | 9 | 8 | 6 | 3 | 1 | 7 | 2 | 4 |
| 1 | 4 | 3 | 7 | 2 | 5 | 8 | 6 | 9 |
| 3 | 1 | 7 | 5 | 4 | 2 | 6 | 9 | 8 |
| 8 | 5 | 9 | 1 | 6 | 7 | 4 | 3 | 2 |
| 6 | 2 | 4 | 3 | 9 | 8 | 1 | 5 | 7 |
| 4 | 8 | 5 | 2 | 1 | 3 | 9 | 7 | 6 |
| 9 | 3 | 1 | 8 | 7 | 6 | 2 | 4 | 5 |
| 7 | 6 | 2 | 9 | 5 | 4 | 3 | 8 | 1 |

## Puzzle 64

| 5 | 3 | 9 | 1 | 2 | 8 | 7 | 6 | 4 |
|---|---|---|---|---|---|---|---|---|
| 1 | 8 | 6 | 9 | 7 | 4 | 2 | 5 | 3 |
| 2 | 4 | 7 | 3 | 6 | 5 | 8 | 1 | 9 |
| 7 | 1 | 5 | 4 | 3 | 2 | 9 | 8 | 6 |
| 3 | 6 | 8 | 5 | 9 | 7 | 1 | 4 | 2 |
| 4 | 9 | 2 | 6 | 8 | 1 | 5 | 3 | 7 |
| 8 | 7 | 4 | 2 | 5 | 6 | 3 | 9 | 1 |
| 6 | 2 | 3 | 8 | 1 | 9 | 4 | 7 | 5 |
| 9 | 5 | 1 | 7 | 4 | 3 | 6 | 2 | 8 |

## Puzzle 65

| 9 | 1 | 6 | 5 | 3 | 8 | 7 | 4 | 2 |
|---|---|---|---|---|---|---|---|---|
| 7 | 5 | 3 | 2 | 4 | 6 | 1 | 8 | 9 |
| 8 | 4 | 2 | 7 | 9 | 1 | 5 | 6 | 3 |
| 5 | 9 | 1 | 6 | 2 | 3 | 4 | 7 | 8 |
| 3 | 8 | 7 | 4 | 5 | 9 | 6 | 2 | 1 |
| 2 | 6 | 4 | 1 | 8 | 7 | 3 | 9 | 5 |
| 6 | 2 | 9 | 3 | 7 | 5 | 8 | 1 | 4 |
| 1 | 3 | 8 | 9 | 6 | 4 | 2 | 5 | 7 |
| 4 | 7 | 5 | 8 | 1 | 2 | 9 | 3 | 6 |

## Puzzle 66

| 3 | 8 | 1 | 5 | 2 | 6 | 4 | 9 | 7 |
|---|---|---|---|---|---|---|---|---|
| 2 | 7 | 6 | 4 | 8 | 9 | 3 | 5 | 1 |
| 5 | 9 | 4 | 1 | 3 | 7 | 8 | 2 | 6 |
| 6 | 4 | 9 | 3 | 1 | 8 | 5 | 7 | 2 |
| 1 | 5 | 8 | 7 | 9 | 2 | 6 | 3 | 4 |
| 7 | 3 | 2 | 6 | 4 | 5 | 1 | 8 | 9 |
| 9 | 1 | 7 | 8 | 6 | 3 | 2 | 4 | 5 |
| 4 | 2 | 3 | 9 | 5 | 1 | 7 | 6 | 8 |
| 8 | 6 | 5 | 2 | 7 | 4 | 9 | 1 | 3 |

## Puzzle 67

| 1 | 8 | 4 | 7 | 2 | 9 | 3 | 6 | 5 |
|---|---|---|---|---|---|---|---|---|
| 7 | 3 | 5 | 8 | 6 | 4 | 2 | 9 | 1 |
| 9 | 6 | 2 | 1 | 3 | 5 | 8 | 7 | 4 |
| 3 | 2 | 6 | 4 | 5 | 1 | 9 | 8 | 7 |
| 5 | 9 | 8 | 2 | 7 | 6 | 1 | 4 | 3 |
| 4 | 7 | 1 | 9 | 8 | 3 | 6 | 5 | 2 |
| 6 | 5 | 7 | 3 | 1 | 8 | 4 | 2 | 9 |
| 8 | 1 | 9 | 5 | 4 | 2 | 7 | 3 | 6 |
| 2 | 4 | 3 | 6 | 9 | 7 | 5 | 1 | 8 |

## Puzzle 68

| 8 | 3 | 1 | 5 | 6 | 2 | 7 | 9 | 4 |
|---|---|---|---|---|---|---|---|---|
| 9 | 7 | 4 | 1 | 3 | 8 | 5 | 2 | 6 |
| 2 | 6 | 5 | 7 | 9 | 4 | 8 | 1 | 3 |
| 3 | 8 | 2 | 4 | 7 | 6 | 9 | 5 | 1 |
| 1 | 5 | 7 | 9 | 8 | 3 | 6 | 4 | 2 |
| 6 | 4 | 9 | 2 | 1 | 5 | 3 | 7 | 8 |
| 7 | 2 | 6 | 8 | 4 | 9 | 1 | 3 | 5 |
| 4 | 1 | 8 | 3 | 5 | 7 | 2 | 6 | 9 |
| 5 | 9 | 3 | 6 | 2 | 1 | 4 | 8 | 7 |

## Puzzle 69

| 5 | 3 | 6 | 4 | 2 | 1 | 9 | 7 | 8 |
|---|---|---|---|---|---|---|---|---|
| 8 | 4 | 1 | 7 | 9 | 3 | 2 | 6 | 5 |
| 9 | 7 | 2 | 8 | 5 | 6 | 1 | 3 | 4 |
| 2 | 6 | 8 | 1 | 4 | 9 | 3 | 5 | 7 |
| 4 | 9 | 7 | 2 | 3 | 5 | 6 | 8 | 1 |
| 3 | 1 | 5 | 6 | 7 | 8 | 4 | 9 | 2 |
| 7 | 8 | 9 | 3 | 1 | 4 | 5 | 2 | 6 |
| 1 | 2 | 3 | 5 | 6 | 7 | 8 | 4 | 9 |
| 6 | 5 | 4 | 9 | 8 | 2 | 7 | 1 | 3 |

## Puzzle 70

| 5 | 4 | 2 | 7 | 9 | 6 | 1 | 8 | 3 |
|---|---|---|---|---|---|---|---|---|
| 8 | 1 | 9 | 3 | 5 | 2 | 4 | 7 | 6 |
| 6 | 3 | 7 | 4 | 8 | 1 | 9 | 2 | 5 |
| 7 | 6 | 4 | 1 | 3 | 9 | 8 | 5 | 2 |
| 1 | 8 | 3 | 2 | 7 | 5 | 6 | 4 | 9 |
| 9 | 2 | 5 | 6 | 4 | 8 | 3 | 1 | 7 |
| 4 | 7 | 6 | 8 | 2 | 3 | 5 | 9 | 1 |
| 2 | 9 | 1 | 5 | 6 | 4 | 7 | 3 | 8 |
| 3 | 5 | 8 | 9 | 1 | 7 | 2 | 6 | 4 |

## Puzzle 71

| 4 | 3 | 9 | 5 | 6 | 2 | 8 | 1 | 7 |
|---|---|---|---|---|---|---|---|---|
| 1 | 5 | 2 | 7 | 3 | 8 | 9 | 6 | 4 |
| 6 | 8 | 7 | 4 | 9 | 1 | 5 | 3 | 2 |
| 3 | 1 | 5 | 6 | 2 | 9 | 4 | 7 | 8 |
| 7 | 9 | 6 | 3 | 8 | 4 | 1 | 2 | 5 |
| 2 | 4 | 8 | 1 | 5 | 7 | 3 | 9 | 6 |
| 9 | 7 | 1 | 2 | 4 | 5 | 6 | 8 | 3 |
| 5 | 2 | 3 | 8 | 1 | 6 | 7 | 4 | 9 |
| 8 | 6 | 4 | 9 | 7 | 3 | 2 | 5 | 1 |

## Puzzle 72

| 2 | 3 | 7 | 4 | 5 | 8 | 1 | 6 | 9 |
|---|---|---|---|---|---|---|---|---|
| 1 | 6 | 5 | 2 | 9 | 7 | 4 | 8 | 3 |
| 9 | 8 | 4 | 1 | 3 | 6 | 2 | 7 | 5 |
| 8 | 1 | 2 | 5 | 6 | 3 | 9 | 4 | 7 |
| 5 | 7 | 6 | 9 | 4 | 2 | 8 | 3 | 1 |
| 3 | 4 | 9 | 7 | 8 | 1 | 6 | 5 | 2 |
| 4 | 5 | 1 | 6 | 7 | 9 | 3 | 2 | 8 |
| 7 | 9 | 8 | 3 | 2 | 4 | 5 | 1 | 6 |
| 6 | 2 | 3 | 8 | 1 | 5 | 7 | 9 | 4 |

## Puzzle 73

| 2 | 5 | 3 | 8 | 6 | 1 | 9 | 7 | 4 |
|---|---|---|---|---|---|---|---|---|
| 7 | 1 | 8 | 4 | 9 | 5 | 2 | 6 | 3 |
| 4 | 9 | 6 | 7 | 2 | 3 | 5 | 1 | 8 |
| 1 | 3 | 2 | 5 | 4 | 8 | 6 | 9 | 7 |
| 6 | 4 | 9 | 2 | 1 | 7 | 8 | 3 | 5 |
| 8 | 7 | 5 | 6 | 3 | 9 | 1 | 4 | 2 |
| 9 | 2 | 4 | 3 | 8 | 6 | 7 | 5 | 1 |
| 3 | 6 | 7 | 1 | 5 | 2 | 4 | 8 | 9 |
| 5 | 8 | 1 | 9 | 7 | 4 | 3 | 2 | 6 |

## Puzzle 74

| 5 | 8 | 9 | 2 | 7 | 1 | 3 | 4 | 6 |
|---|---|---|---|---|---|---|---|---|
| 2 | 1 | 3 | 6 | 4 | 5 | 7 | 9 | 8 |
| 6 | 7 | 4 | 3 | 9 | 8 | 2 | 1 | 5 |
| 4 | 9 | 5 | 7 | 6 | 2 | 1 | 8 | 3 |
| 8 | 6 | 1 | 9 | 5 | 3 | 4 | 7 | 2 |
| 3 | 2 | 7 | 1 | 8 | 4 | 5 | 6 | 9 |
| 1 | 5 | 8 | 4 | 2 | 9 | 6 | 3 | 7 |
| 7 | 4 | 2 | 8 | 3 | 6 | 9 | 5 | 1 |
| 9 | 3 | 6 | 5 | 1 | 7 | 8 | 2 | 4 |

## Puzzle 75

| 2 | 6 | 7 | 4 | 3 | 9 | 8 | 1 | 5 |
|---|---|---|---|---|---|---|---|---|
| 4 | 8 | 1 | 5 | 6 | 7 | 2 | 3 | 9 |
| 9 | 5 | 3 | 1 | 2 | 8 | 4 | 7 | 6 |
| 8 | 1 | 5 | 7 | 4 | 6 | 3 | 9 | 2 |
| 3 | 7 | 4 | 2 | 9 | 1 | 5 | 6 | 8 |
| 6 | 2 | 9 | 8 | 5 | 3 | 1 | 4 | 7 |
| 7 | 3 | 8 | 6 | 1 | 2 | 9 | 5 | 4 |
| 1 | 4 | 6 | 9 | 8 | 5 | 7 | 2 | 3 |
| 5 | 9 | 2 | 3 | 7 | 4 | 6 | 8 | 1 |

## Puzzle 76

| 3 | 9 | 4 | 1 | 2 | 6 | 7 | 8 | 5 |
|---|---|---|---|---|---|---|---|---|
| 1 | 6 | 7 | 4 | 8 | 5 | 2 | 9 | 3 |
| 5 | 2 | 8 | 3 | 9 | 7 | 1 | 6 | 4 |
| 8 | 4 | 1 | 2 | 5 | 3 | 9 | 7 | 6 |
| 7 | 3 | 6 | 8 | 1 | 9 | 4 | 5 | 2 |
| 2 | 5 | 9 | 7 | 6 | 4 | 3 | 1 | 8 |
| 6 | 1 | 2 | 9 | 4 | 8 | 5 | 3 | 7 |
| 4 | 7 | 5 | 6 | 3 | 1 | 8 | 2 | 9 |
| 9 | 8 | 3 | 5 | 7 | 2 | 6 | 4 | 1 |

## Puzzle 77

| 8 | 1 | 9 | 4 | 7 | 5 | 6 | 3 | 2 |
|---|---|---|---|---|---|---|---|---|
| 4 | 3 | 7 | 6 | 8 | 2 | 1 | 5 | 9 |
| 5 | 2 | 6 | 3 | 9 | 1 | 4 | 8 | 7 |
| 6 | 5 | 8 | 9 | 4 | 7 | 3 | 2 | 1 |
| 7 | 9 | 1 | 5 | 2 | 3 | 8 | 6 | 4 |
| 3 | 4 | 2 | 8 | 1 | 6 | 7 | 9 | 5 |
| 9 | 8 | 5 | 1 | 6 | 4 | 2 | 7 | 3 |
| 2 | 6 | 4 | 7 | 3 | 9 | 5 | 1 | 8 |
| 1 | 7 | 3 | 2 | 5 | 8 | 9 | 4 | 6 |

## Puzzle 78

| 7 | 1 | 4 | 5 | 6 | 3 | 9 | 2 | 8 |
|---|---|---|---|---|---|---|---|---|
| 3 | 6 | 2 | 1 | 9 | 8 | 5 | 4 | 7 |
| 9 | 5 | 8 | 4 | 2 | 7 | 6 | 1 | 3 |
| 5 | 2 | 1 | 8 | 3 | 6 | 4 | 7 | 9 |
| 6 | 3 | 7 | 9 | 1 | 4 | 2 | 8 | 5 |
| 4 | 8 | 9 | 2 | 7 | 5 | 1 | 3 | 6 |
| 1 | 9 | 3 | 6 | 8 | 2 | 7 | 5 | 4 |
| 2 | 7 | 5 | 3 | 4 | 9 | 8 | 6 | 1 |
| 8 | 4 | 6 | 7 | 5 | 1 | 3 | 9 | 2 |

## Puzzle 79

| 7 | 4 | 2 | 1 | 8 | 3 | 5 | 6 | 9 |
|---|---|---|---|---|---|---|---|---|
| 3 | 5 | 1 | 9 | 7 | 6 | 2 | 8 | 4 |
| 9 | 8 | 6 | 2 | 5 | 4 | 3 | 1 | 7 |
| 6 | 1 | 5 | 4 | 3 | 7 | 9 | 2 | 8 |
| 4 | 2 | 3 | 8 | 9 | 1 | 7 | 5 | 6 |
| 8 | 7 | 9 | 5 | 6 | 2 | 4 | 3 | 1 |
| 1 | 9 | 8 | 7 | 2 | 5 | 6 | 4 | 3 |
| 2 | 6 | 7 | 3 | 4 | 8 | 1 | 9 | 5 |
| 5 | 3 | 4 | 6 | 1 | 9 | 8 | 7 | 2 |

## Puzzle 80

| 3 | 4 | 5 | 2 | 1 | 9 | 6 | 8 | 7 |
|---|---|---|---|---|---|---|---|---|
| 7 | 2 | 1 | 3 | 6 | 8 | 9 | 4 | 5 |
| 9 | 8 | 6 | 4 | 7 | 5 | 2 | 3 | 1 |
| 1 | 6 | 8 | 7 | 2 | 3 | 5 | 9 | 4 |
| 5 | 3 | 4 | 6 | 9 | 1 | 8 | 7 | 2 |
| 2 | 9 | 7 | 8 | 5 | 4 | 1 | 6 | 3 |
| 4 | 5 | 3 | 1 | 8 | 6 | 7 | 2 | 9 |
| 6 | 1 | 2 | 9 | 3 | 7 | 4 | 5 | 8 |
| 8 | 7 | 9 | 5 | 4 | 2 | 3 | 1 | 6 |

## Puzzle 81

| 3 | 7 | 2 | 6 | 4 | 8 | 5 | 9 | 1 |
|---|---|---|---|---|---|---|---|---|
| 9 | 1 | 4 | 3 | 7 | 5 | 2 | 8 | 6 |
| 8 | 5 | 6 | 9 | 1 | 2 | 7 | 4 | 3 |
| 7 | 9 | 5 | 4 | 8 | 1 | 3 | 6 | 2 |
| 2 | 4 | 8 | 7 | 3 | 6 | 9 | 1 | 5 |
| 1 | 6 | 3 | 2 | 5 | 9 | 8 | 7 | 4 |
| 4 | 3 | 1 | 5 | 9 | 7 | 6 | 2 | 8 |
| 6 | 8 | 7 | 1 | 2 | 3 | 4 | 5 | 9 |
| 5 | 2 | 9 | 8 | 6 | 4 | 1 | 3 | 7 |

## Puzzle 82

| 2 | 1 | 3 | 4 | 7 | 8 | 5 | 6 | 9 |
|---|---|---|---|---|---|---|---|---|
| 7 | 6 | 9 | 2 | 5 | 3 | 4 | 8 | 1 |
| 4 | 5 | 8 | 1 | 9 | 6 | 2 | 3 | 7 |
| 8 | 2 | 5 | 6 | 1 | 7 | 9 | 4 | 3 |
| 6 | 9 | 4 | 3 | 8 | 2 | 7 | 1 | 5 |
| 1 | 3 | 7 | 5 | 4 | 9 | 8 | 2 | 6 |
| 3 | 7 | 6 | 8 | 2 | 5 | 1 | 9 | 4 |
| 5 | 8 | 1 | 9 | 3 | 4 | 6 | 7 | 2 |
| 9 | 4 | 2 | 7 | 6 | 1 | 3 | 5 | 8 |

## Puzzle 83

| 4 | 5 | 2 | 8 | 1 | 6 | 7 | 9 | 3 |
|---|---|---|---|---|---|---|---|---|
| 6 | 7 | 9 | 3 | 5 | 4 | 8 | 1 | 2 |
| 3 | 1 | 8 | 2 | 9 | 7 | 6 | 5 | 4 |
| 5 | 8 | 1 | 6 | 3 | 9 | 4 | 2 | 7 |
| 2 | 6 | 3 | 7 | 4 | 1 | 5 | 8 | 9 |
| 7 | 9 | 4 | 5 | 8 | 2 | 3 | 6 | 1 |
| 8 | 2 | 7 | 9 | 6 | 3 | 1 | 4 | 5 |
| 1 | 3 | 6 | 4 | 2 | 5 | 9 | 7 | 8 |
| 9 | 4 | 5 | 1 | 7 | 8 | 2 | 3 | 6 |

## Puzzle 84

| 5 | 1 | 9 | 3 | 7 | 2 | 6 | 8 | 4 |
|---|---|---|---|---|---|---|---|---|
| 7 | 8 | 6 | 4 | 1 | 9 | 2 | 5 | 3 |
| 3 | 4 | 2 | 6 | 5 | 8 | 7 | 1 | 9 |
| 4 | 6 | 8 | 7 | 2 | 5 | 9 | 3 | 1 |
| 2 | 9 | 5 | 1 | 8 | 3 | 4 | 6 | 7 |
| 1 | 3 | 7 | 9 | 4 | 6 | 8 | 2 | 5 |
| 6 | 7 | 1 | 8 | 3 | 4 | 5 | 9 | 2 |
| 9 | 2 | 3 | 5 | 6 | 7 | 1 | 4 | 8 |
| 8 | 5 | 4 | 2 | 9 | 1 | 3 | 7 | 6 |

## Puzzle 85

| 3 | 4 | 9 | 2 | 5 | 7 | 8 | 1 | 6 |
| 1 | 7 | 5 | 6 | 8 | 4 | 9 | 3 | 2 |
| 2 | 8 | 6 | 3 | 1 | 9 | 7 | 4 | 5 |
| 9 | 2 | 4 | 7 | 3 | 1 | 6 | 5 | 8 |
| 5 | 6 | 3 | 4 | 9 | 8 | 2 | 7 | 1 |
| 8 | 1 | 7 | 5 | 2 | 6 | 4 | 9 | 3 |
| 7 | 5 | 1 | 8 | 4 | 2 | 3 | 6 | 9 |
| 4 | 9 | 8 | 1 | 6 | 3 | 5 | 2 | 7 |
| 6 | 3 | 2 | 9 | 7 | 5 | 1 | 8 | 4 |

## Puzzle 86

| 1 | 3 | 2 | 9 | 7 | 8 | 5 | 4 | 6 |
| 6 | 9 | 7 | 5 | 2 | 4 | 3 | 1 | 8 |
| 5 | 8 | 4 | 1 | 6 | 3 | 9 | 7 | 2 |
| 7 | 1 | 8 | 3 | 5 | 6 | 2 | 9 | 4 |
| 3 | 5 | 9 | 7 | 4 | 2 | 8 | 6 | 1 |
| 2 | 4 | 6 | 8 | 1 | 9 | 7 | 5 | 3 |
| 8 | 7 | 3 | 6 | 9 | 1 | 4 | 2 | 5 |
| 4 | 6 | 5 | 2 | 8 | 7 | 1 | 3 | 9 |
| 9 | 2 | 1 | 4 | 3 | 5 | 6 | 8 | 7 |

## Puzzle 87

| 2 | 8 | 3 | 5 | 6 | 1 | 4 | 9 | 7 |
| 4 | 7 | 9 | 8 | 3 | 2 | 1 | 5 | 6 |
| 6 | 1 | 5 | 4 | 9 | 7 | 3 | 8 | 2 |
| 3 | 6 | 2 | 9 | 1 | 5 | 7 | 4 | 8 |
| 9 | 4 | 7 | 3 | 8 | 6 | 5 | 2 | 1 |
| 8 | 5 | 1 | 2 | 7 | 4 | 6 | 3 | 9 |
| 1 | 2 | 4 | 7 | 5 | 8 | 9 | 6 | 3 |
| 7 | 9 | 8 | 6 | 4 | 3 | 2 | 1 | 5 |
| 5 | 3 | 6 | 1 | 2 | 9 | 8 | 7 | 4 |

## Puzzle 88

| 3 | 7 | 4 | 9 | 5 | 6 | 8 | 2 | 1 |
| 5 | 1 | 9 | 7 | 2 | 8 | 4 | 6 | 3 |
| 6 | 8 | 2 | 3 | 4 | 1 | 9 | 7 | 5 |
| 9 | 5 | 3 | 1 | 6 | 7 | 2 | 8 | 4 |
| 4 | 6 | 1 | 8 | 3 | 2 | 7 | 5 | 9 |
| 8 | 2 | 7 | 5 | 9 | 4 | 1 | 3 | 6 |
| 2 | 9 | 6 | 4 | 8 | 5 | 3 | 1 | 7 |
| 7 | 3 | 5 | 2 | 1 | 9 | 6 | 4 | 8 |
| 1 | 4 | 8 | 6 | 7 | 3 | 5 | 9 | 2 |

## Puzzle 89

| 9 | 7 | 8 | 4 | 3 | 1 | 6 | 2 | 5 |
| 3 | 5 | 1 | 8 | 2 | 6 | 4 | 9 | 7 |
| 4 | 6 | 2 | 9 | 5 | 7 | 8 | 1 | 3 |
| 2 | 1 | 6 | 5 | 4 | 8 | 7 | 3 | 9 |
| 7 | 4 | 3 | 6 | 9 | 2 | 1 | 5 | 8 |
| 8 | 9 | 5 | 7 | 1 | 3 | 2 | 6 | 4 |
| 5 | 8 | 4 | 1 | 6 | 9 | 3 | 7 | 2 |
| 1 | 3 | 9 | 2 | 7 | 4 | 5 | 8 | 6 |
| 6 | 2 | 7 | 3 | 8 | 5 | 9 | 4 | 1 |

## Puzzle 90

| 7 | 8 | 6 | 4 | 2 | 3 | 5 | 1 | 9 |
| 5 | 1 | 3 | 7 | 6 | 9 | 4 | 2 | 8 |
| 9 | 2 | 4 | 1 | 8 | 5 | 6 | 7 | 3 |
| 3 | 9 | 7 | 2 | 5 | 4 | 1 | 8 | 6 |
| 1 | 4 | 5 | 6 | 9 | 8 | 7 | 3 | 2 |
| 8 | 6 | 2 | 3 | 1 | 7 | 9 | 4 | 5 |
| 2 | 7 | 9 | 8 | 4 | 6 | 3 | 5 | 1 |
| 6 | 3 | 1 | 5 | 7 | 2 | 8 | 9 | 4 |
| 4 | 5 | 8 | 9 | 3 | 1 | 2 | 6 | 7 |

## Puzzle 91

| 1 | 3 | 8 | 2 | 6 | 4 | 5 | 7 | 9 |
|---|---|---|---|---|---|---|---|---|
| 9 | 5 | 6 | 7 | 8 | 1 | 4 | 2 | 3 |
| 2 | 7 | 4 | 5 | 3 | 9 | 1 | 6 | 8 |
| 7 | 4 | 5 | 3 | 9 | 8 | 2 | 1 | 6 |
| 3 | 6 | 2 | 1 | 7 | 5 | 8 | 9 | 4 |
| 8 | 9 | 1 | 4 | 2 | 6 | 3 | 5 | 7 |
| 5 | 8 | 7 | 6 | 1 | 3 | 9 | 4 | 2 |
| 6 | 1 | 3 | 9 | 4 | 2 | 7 | 8 | 5 |
| 4 | 2 | 9 | 8 | 5 | 7 | 6 | 3 | 1 |

## Puzzle 92

| 2 | 7 | 4 | 6 | 9 | 1 | 3 | 8 | 5 |
|---|---|---|---|---|---|---|---|---|
| 6 | 5 | 3 | 7 | 2 | 8 | 1 | 4 | 9 |
| 1 | 8 | 9 | 3 | 4 | 5 | 7 | 6 | 2 |
| 9 | 4 | 7 | 5 | 3 | 6 | 8 | 2 | 1 |
| 5 | 6 | 2 | 1 | 8 | 9 | 4 | 3 | 7 |
| 8 | 3 | 1 | 4 | 7 | 2 | 9 | 5 | 6 |
| 4 | 2 | 8 | 9 | 5 | 7 | 6 | 1 | 3 |
| 7 | 1 | 5 | 8 | 6 | 3 | 2 | 9 | 4 |
| 3 | 9 | 6 | 2 | 1 | 4 | 5 | 7 | 8 |

## Puzzle 93

| 4 | 6 | 7 | 9 | 1 | 5 | 8 | 2 | 3 |
|---|---|---|---|---|---|---|---|---|
| 5 | 3 | 9 | 2 | 7 | 8 | 4 | 1 | 6 |
| 2 | 1 | 8 | 4 | 3 | 6 | 5 | 7 | 9 |
| 8 | 4 | 5 | 3 | 9 | 7 | 2 | 6 | 1 |
| 6 | 2 | 3 | 1 | 8 | 4 | 9 | 5 | 7 |
| 7 | 9 | 1 | 5 | 6 | 2 | 3 | 4 | 8 |
| 1 | 8 | 2 | 7 | 4 | 3 | 6 | 9 | 5 |
| 3 | 7 | 4 | 6 | 5 | 9 | 1 | 8 | 2 |
| 9 | 5 | 6 | 8 | 2 | 1 | 7 | 3 | 4 |

## Puzzle 94

| 7 | 5 | 4 | 1 | 9 | 3 | 2 | 6 | 8 |
|---|---|---|---|---|---|---|---|---|
| 3 | 8 | 9 | 2 | 5 | 6 | 7 | 4 | 1 |
| 6 | 2 | 1 | 8 | 4 | 7 | 3 | 5 | 9 |
| 5 | 9 | 7 | 3 | 1 | 4 | 8 | 2 | 6 |
| 4 | 1 | 6 | 9 | 2 | 8 | 5 | 7 | 3 |
| 2 | 3 | 8 | 7 | 6 | 5 | 1 | 9 | 4 |
| 1 | 6 | 2 | 5 | 8 | 9 | 4 | 3 | 7 |
| 8 | 4 | 3 | 6 | 7 | 2 | 9 | 1 | 5 |
| 9 | 7 | 5 | 4 | 3 | 1 | 6 | 8 | 2 |

## Puzzle 95

| 4 | 8 | 5 | 6 | 2 | 1 | 9 | 3 | 7 |
|---|---|---|---|---|---|---|---|---|
| 6 | 9 | 2 | 7 | 3 | 5 | 8 | 4 | 1 |
| 3 | 7 | 1 | 8 | 4 | 9 | 2 | 6 | 5 |
| 8 | 4 | 7 | 5 | 6 | 3 | 1 | 2 | 9 |
| 1 | 2 | 6 | 9 | 8 | 4 | 5 | 7 | 3 |
| 9 | 5 | 3 | 1 | 7 | 2 | 6 | 8 | 4 |
| 2 | 1 | 8 | 4 | 9 | 7 | 3 | 5 | 6 |
| 5 | 3 | 4 | 2 | 1 | 6 | 7 | 9 | 8 |
| 7 | 6 | 9 | 3 | 5 | 8 | 4 | 1 | 2 |

## Puzzle 96

| 1 | 3 | 6 | 8 | 4 | 7 | 5 | 9 | 2 |
|---|---|---|---|---|---|---|---|---|
| 4 | 2 | 7 | 5 | 3 | 9 | 6 | 1 | 8 |
| 5 | 9 | 8 | 1 | 2 | 6 | 3 | 7 | 4 |
| 7 | 1 | 3 | 6 | 5 | 8 | 2 | 4 | 9 |
| 9 | 8 | 4 | 2 | 1 | 3 | 7 | 5 | 6 |
| 2 | 6 | 5 | 7 | 9 | 4 | 1 | 8 | 3 |
| 3 | 4 | 1 | 9 | 7 | 2 | 8 | 6 | 5 |
| 6 | 7 | 2 | 4 | 8 | 5 | 9 | 3 | 1 |
| 8 | 5 | 9 | 3 | 6 | 1 | 4 | 2 | 7 |

## Puzzle 97

| 4 | 7 | 6 | 1 | 8 | 5 | 9 | 3 | 2 |
|---|---|---|---|---|---|---|---|---|
| 3 | 5 | 2 | 4 | 9 | 6 | 8 | 1 | 7 |
| 8 | 1 | 9 | 7 | 2 | 3 | 5 | 4 | 6 |
| 6 | 9 | 3 | 8 | 4 | 7 | 2 | 5 | 1 |
| 2 | 4 | 5 | 3 | 6 | 1 | 7 | 9 | 8 |
| 1 | 8 | 7 | 2 | 5 | 9 | 4 | 6 | 3 |
| 5 | 3 | 1 | 9 | 7 | 2 | 6 | 8 | 4 |
| 7 | 6 | 4 | 5 | 1 | 8 | 3 | 2 | 9 |
| 9 | 2 | 8 | 6 | 3 | 4 | 1 | 7 | 5 |

## Puzzle 98

| 9 | 6 | 5 | 2 | 3 | 8 | 4 | 1 | 7 |
|---|---|---|---|---|---|---|---|---|
| 3 | 7 | 2 | 1 | 5 | 4 | 9 | 8 | 6 |
| 1 | 4 | 8 | 9 | 6 | 7 | 5 | 2 | 3 |
| 5 | 1 | 9 | 3 | 2 | 6 | 7 | 4 | 8 |
| 4 | 3 | 7 | 8 | 9 | 1 | 6 | 5 | 2 |
| 8 | 2 | 6 | 4 | 7 | 5 | 1 | 3 | 9 |
| 7 | 5 | 1 | 6 | 8 | 2 | 3 | 9 | 4 |
| 6 | 8 | 3 | 5 | 4 | 9 | 2 | 7 | 1 |
| 2 | 9 | 4 | 7 | 1 | 3 | 8 | 6 | 5 |

## Puzzle 99

| 4 | 7 | 2 | 3 | 9 | 1 | 6 | 5 | 8 |
|---|---|---|---|---|---|---|---|---|
| 1 | 5 | 8 | 2 | 6 | 4 | 9 | 3 | 7 |
| 3 | 9 | 6 | 5 | 8 | 7 | 1 | 2 | 4 |
| 6 | 3 | 1 | 8 | 4 | 2 | 5 | 7 | 9 |
| 5 | 2 | 4 | 7 | 1 | 9 | 3 | 8 | 6 |
| 9 | 8 | 7 | 6 | 3 | 5 | 4 | 1 | 2 |
| 8 | 1 | 5 | 9 | 7 | 6 | 2 | 4 | 3 |
| 7 | 4 | 9 | 1 | 2 | 3 | 8 | 6 | 5 |
| 2 | 6 | 3 | 4 | 5 | 8 | 7 | 9 | 1 |

## Puzzle 100

| 2 | 5 | 3 | 8 | 1 | 4 | 7 | 6 | 9 |
|---|---|---|---|---|---|---|---|---|
| 8 | 7 | 6 | 9 | 5 | 2 | 4 | 3 | 1 |
| 4 | 1 | 9 | 7 | 3 | 6 | 5 | 2 | 8 |
| 5 | 3 | 4 | 2 | 7 | 1 | 9 | 8 | 6 |
| 9 | 6 | 2 | 5 | 4 | 8 | 3 | 1 | 7 |
| 1 | 8 | 7 | 6 | 9 | 3 | 2 | 4 | 5 |
| 6 | 9 | 8 | 4 | 2 | 7 | 1 | 5 | 3 |
| 7 | 2 | 1 | 3 | 8 | 5 | 6 | 9 | 4 |
| 3 | 4 | 5 | 1 | 6 | 9 | 8 | 7 | 2 |

www.ingramcontent.com/pod-product-compliance
Lightning Source LLC
Chambersburg PA
CBHW070819220526
45466CB00002B/718